Station Act

for Common Core Mathematics
Grade 6

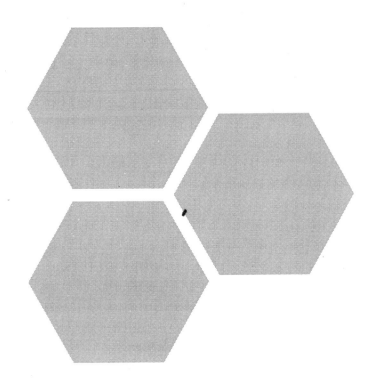

WALCH ☛ EDUCATION®

1   2   3   4   5   6   7   8   9   10

ISBN 978-0-8251-6783-6

Copyright © 2011

J. Weston Walch, Publisher

Portland, ME 04103

www.walch.com

Printed in the United States of America

Table of Contents

Standards Correlations

The standards correlations below support the implementation of the Common Core Standards. This book includes station activity sets for the Common Core domains of Ratios and Proportional Relationships; The Number System; Expressions and Equations; Geometry; and Statistics and Probability. This table provides a listing of the available station activities organized by Common Core standard.

The left column lists the standard codes. The first number of the code represents the grade level. The grade number is followed by the initials of the Common Core domain name, which is then followed by the standard number. The middle column of the table lists the title of the station activity set that corresponds to the standard(s), and the right column lists the page number where the station activity set can be found.

Standard	Set title	Page number
6.RP.1.	Ratio, Proportion, and Scale	1
6.RP.1.	Patterns and Relationships	8
6.RP.1.	Proportional Relationships	15
6.RP.2.	Proportional Relationships	15
6.RP.3	Ratio, Proportion, and Scale	1
6.RP.3.	Patterns and Relationships	8
6.RP.3.	Proportional Relationships	15
6.NS.1.	Multiplying and Dividing Fractions	44
6.NS.4.	Factors, Multiples, and Prime Factorization	37
6.NS.5.	Integers and Absolute Value	22
6.NS.5.	Comparing and Ordering Rational Numbers	29
6.NS.6.	Integers and Absolute Value	22
6.NS.6.	Comparing and Ordering Rational Numbers	29
6.NS.7.	Integers and Absolute Value	22
6.NS.7.	Comparing and Ordering Rational Numbers	29
6.EE.2.	Graphing Relationships	51
6.EE.2.	Evaluating Expressions	60
6.EE.2.	Solving Equations	68
6.EE.3.	Evaluating and Simplifying Expressions	75
6.EE.4.	Evaluating and Simplifying Expressions	75

(continued)

Standards Correlations

Standard	Set title	Page number
6.EE.5.	Evaluating Expressions	60
6.EE.5.	Solving Equations	68
6.EE.6.	Solving Equations	68
6.EE.8.	Solving Inequalities	82
6.EE.9	Graphing Relationships	51
6.G.1.	Problem Solving with Volume, Surface Area, and Scale	106
6.G.2.	Appropriate Units of Measurement	91
6.G.2.	Problem Solving with Volume, Surface Area, and Scale	106
6.G.4.	Visualizing Solid Figures	98
6.SP.1.	Collecting, Organizing, and Analyzing Data	113
6.SP.2.	Collecting, Organizing, and Analyzing Data	113
6.SP.2.	Measures of Central Tendency	135
6.SP.2.	Measures of Variation	143
6.SP.3.	Collecting, Organizing, and Analyzing Data	113
6.SP.3.	Measures of Central Tendency	135
6.SP.3.	Measures of Variation	143
6.SP.4.	Constructing Frequency Distributions	120
6.SP.4.	Using Tables and Graphs	127
6.SP.4.	Analyzing Data Using Graphs	150
6.SP.5.	Constructing Frequency Distributions	120
6.SP.5.	Using Tables and Graphs	127
6.SP.5.	Analyzing Data Using Graphs	150

Introduction

This book includes a collection of station-based activities to provide students with opportunities to practice and apply the mathematical skills and concepts they are learning. It contains sets of activities for each of the five Grade 6 Common Core Mathematics strands: Ratios and Proportional Relationships; The Number System; Expressions and Equations; Geometry; and Statistics and Probability. You may use these activities in addition to direct instruction, or instead of direct instruction in areas where students understand the basic concepts but need practice. The Discussion Guide included with each set of activities provides an important opportunity to help students reflect on their experiences and synthesize their thinking. It also provides guidance for ongoing, informal assessment to inform instructional planning.

Implementation Guide

The following guidelines will help you prepare for and use the activity sets in this book.

Setting Up the Stations

Each activity set consists of four stations. Set up each station at a desk, or at several desks pushed together, with enough chairs for a small group of students. Place a card with the number of the station on the desk. Each station should also contain the materials specified in the teacher's notes, and a stack of student activity sheets (one copy per student). Place the required materials (as listed) at each station.

When a group of students arrives at a station, each student should take one of the activity sheets to record the group's work. Although students should work together to develop one set of answers for the entire group, each student should record the answers on his or her own activity sheet. This helps keep students engaged in the activity and gives each student a record of the activity for future reference.

Forming Groups of Students

All activity sets consist of four stations. You might divide the class into four groups by having students count off from 1 to 4. If you have a large class and want to have students working in small groups, you might set up two identical sets of stations, labeled A and B. In this way, the class can be divided into eight groups, with each group of students rotating through the "A" stations or "B" stations.

Assigning Roles to Students

Students often work most productively in groups when each student has an assigned role. You may want to assign roles to students when they are assigned to groups and change the roles occasionally. Some possible roles are as follows:

- Reader—reads the steps of the activity aloud
- Facilitator—makes sure that each student in the group has a chance to speak and pose questions; also makes sure that each student agrees on each answer before it is written down
- Materials Manager—handles the materials at the station and makes sure the materials are put back in place at the end of the activity
- Timekeeper—tracks the group's progress to ensure that the activity is completed in the allotted time
- Spokesperson—speaks for the group during the debriefing session after the activities

Timing the Activities

The activities in this book are designed to take approximately 15 minutes per station. Therefore, you might plan on having groups change stations every 15 minutes, with a two-minute interval for moving from one station to the next. It is helpful to give students a "5-minute warning" before it is time to change stations.

Since the activity sets consist of four stations, the above time frame means that it will take about an hour and 10 minutes for groups to work through all stations. If this is followed by a 20-minute class discussion as described below, an entire activity set can be completed in about 90 minutes.

Guidelines for Students

Before starting the first activity set, you may want to review the following "ground rules" with students. You might also post the rules in the classroom.

- All students in a group should agree on each answer before it is written down. If there is a disagreement within the group, discuss it with one another.
- You can ask your teacher a question only if everyone in the group has the same question.
- If you finish early, work together to write problems of your own that are similar to the ones on the student activity sheet.
- Leave the station exactly as you found it. All materials should be in the same place and in the same condition as when you arrived.

Debriefing the Activities

After each group has rotated through every station, bring students together for a brief class discussion. At this time, you might have the groups' spokespersons pose any questions they had about the activities. Before responding, ask if students in other groups encountered the same difficulty or if they have a response to the question. The class discussion is also a good time to reinforce the essential ideas of the activities. The questions that are provided in the teacher's notes for each activity set can serve as a guide to initiating this type of discussion.

You may want to collect the student activity sheets before beginning the class discussion. However, it can be beneficial to collect the sheets afterward so that students can refer to them during the discussion. This also gives students a chance to revisit and refine their work based on the debriefing session.

Materials List

Class Sets

- calculators
- rulers

Station Sets

- 3 different figures made up of 10 or fewer connecting cubes
- 3 paper cups
- 12 counters or other small objects (e.g., pennies, beans)
- 25 connecting cubes
- bags of fun-size M&Ms® (plain); one per group member
- bags of fun-size Skittles® (original fruit flavor); one per group member
- box of toothpicks
- equation mat
- graph paper
- highlighters (1 each of yellow and blue)
- mini marshmallows
- newspaper article—150 words or less in length
- pennies (4)
- 8 ½" × 11" paper (several sheets)
- protractors
- rectangular strips of paper or adding machine tape, about 1 inch wide and 1 foot long
- red and blue pencils
- small picture and a larger picture frame (must be similar)
- tape
- tiles or small pieces of paper (several small square tiles; several small round tiles; 12 small red tiles; 18 small blue tiles)
- variety of items that are rectangular prisms
- yard stick

Ongoing Use

- index cards (prepared according to specifications in teacher notes for many of the station activities)
- number cube (numbered 1–6)
- pencils
- pennies

Overhead Manipulatives (optional)

- clock
- spinners
- tiles

Ratios and Proportional Relationships

Set 1: Ratio, Proportion, and Scale

Goal: To provide opportunities for students to develop concepts and skills related to demonstrating the relationship between similar plane figures using ratio, proportion, and scale factor

Common Core Standards

Ratios and Proportional Relationships

Understand ratio concepts and use ratio reasoning to solve problems.

6.RP.1. Understand the concept of a ratio and use ratio language to describe a ratio relationship between two quantities.

6.RP.3. Use ratio and rate reasoning to solve real-world and mathematical problems, e.g., by reasoning about tables of equivalent ratios, tape diagrams, double number line diagrams, or equations.

Student Activities Overview and Answer Key

Station 1

Students construct their own triangles with given angle measurements. They then compare their triangles and explore what similarity means with respect to triangles.

Answers

The ratios are the same; yes, because the ratios are the same and the angles are the same

Station 2

Students will use a drawing of a room and the scale to determine the dimensions of an actual room. They then explain their strategy for successfully completing the activity.

Answers

$1 \frac{9}{16}$ in $\times$ $2 \frac{1}{8}$ in; 15.625 ft $\times$ 21.25 ft; many possible answers—set up a proportion, etc.

Station 3

Students apply scale factor and draw a larger picture using this concept. They then reflect on their strategies for successfully completing the task.

Answers

Answers will vary—setting up proportions, etc.

© 2011 Walch Education

Station 4

Students have two quadrilaterals that they work with to determine scale factor. They measure each corresponding side and find the ratio. They then reflect on the task.

Answers

$5/4$; that is the ratio you need to multiply the length of the sides of the smaller quadrilateral by in order to get the larger quadrilateral

Materials List/Setup

Station 1   pencils, protractors, rulers, and calculators for all group members

Station 2   rulers and calculators for all group members

Station 3   protractors, rulers, and calculators for all group members

Station 4   protractors, rulers, and calculators for all group members

Discussion Guide

To support students in reflecting on the activities, and to gather formative information about student learning, use the following prompts to facilitate a class discussion to "debrief" the station activities.

Prompts/Questions

1. If you had a figure and wanted to construct a similar figure, what information would be important to know?

2. When is scale factor important in real life?

3. How do proportions help you when working with scale factor?

4. When do we use similar figures in real life?

Think, Pair, Share

Have students jot down their own responses to questions, discuss their responses with a partner (who was not in their station group), and then discuss as a whole class.

Suggested Appropriate Responses

1. the angles, proportion of the sides

2. Many possibilities—in blueprints

3. We can set up a proportion to find the new dimensions if we know the scale factor, or we can set up a proportion to find the scale factor if we know our new dimensions.

4. Many possibilities—different size plates—dinner plate vs. bread plate, etc.

Possible Misunderstandings/Mistakes

- Not accurately measuring angles, which will change the ratios
- Not accurately measuring with a ruler, which will change the ratios
- Getting caught up in what the state of Georgia looks like when the concept of scale factor is what is important

Ratios and Proportional Relationships
Set 1: Ratio, Proportion, and Scale

Station 1

At this station you will find rulers, pencils, calculators, and protractors. You will be using these materials to draw a triangle and make observations.

Each group member should draw a triangle in the space below. The angles of the triangle should be 90°, 30°, and 60°. The 90° angle is angle *A*; the 30° angle is angle *B*; the 60° angle is angle *C*.

Each person should measure the sides of his/her triangle. Compile your data in the table below.

Group Member	Length *AB* (cm)	Length *BC* (cm)	Length *AC* (cm)	*AB/BC*	*BC/AC*	*AB/AC*

What do you notice about the ratio of the sides? _____

_____

_____

Are the triangles similar? How do you know? _____

_____

_____

Station Activities for Common Core Mathematics, Grade 6

Ratios and Proportional Relationships

Set 1: Ratio, Proportion, and Scale

Station 2

At this station, you will take on the role of an architect. You will find rulers and calculators to help you with this role.

Below is a diagram of a room in a house. You need to figure out the dimensions of the actual room in the house. For every inch in the diagram, the real room is 10 feet.

Hint: You should use your ruler to find the dimensions of the diagram.

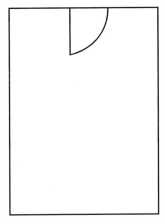

What are the dimensions? _____

What are the dimensions of the actual room? _____

What was your strategy for figuring out the actual dimensions? _____

_____

_____

_____

Ratios and Proportional Relationships
Set 1: Ratio, Proportion, and Scale

Station 3

At this station, you will find enough rulers and calculators for all group members.

Your job is to redraw a picture of the state of Georgia at a scale factor 3 times larger than what it is now.

What was your strategy for successfully completing this task? _____

_____

_____

_____

Ratios and Proportional Relationships
Set 1: Ratio, Proportion, and Scale

Station 4

At this station, you will find rulers and calculators to help you find the scale factor between two figures.

Look at the two figures below. They are similar. Your job is to determine the scale factor used to go from the first figure to the second figure.

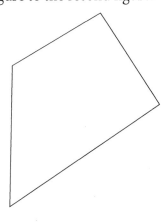 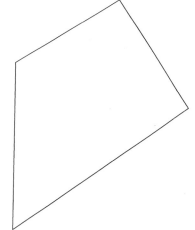

First, measure the corresponding sides.

Side	Quadrilateral 1	Quadrilateral 2
1		
2		
3		
4		

Use this information to determine the scale factor.

What is the scale factor? _____

How do you know? _____

_____

_____

Ratios and Proportional Relationships

Set 2: Patterns and Relationships

Goal: To provide opportunities for students to develop concepts and skills related to patterns and relationships

Common Core Standards

Ratios and Proportional Relationships

Understand ratio concepts and use ratio reasoning to solve problems.

6.RP.1. Understand the concept of a ratio and use ratio language to describe a ratio relationship between two quantities.

6.RP.3. Use ratio and rate reasoning to solve real-world and mathematical problems, e.g., by reasoning about tables of equivalent ratios, tape diagrams, double number line diagrams, or equations.

    a. . . . Use tables to compare ratios.

Student Activities Overview and Answer Key

Station 1

Students build, extend, and describe a geometric pattern made of toothpicks. They are asked to work together to predict the number of toothpicks that would be needed to build the tenth stage of the pattern, and they are asked to describe a general rule for the pattern.

Answers

1. 9, 11; 2. 21; 3. Possible answer: Double the number of the stage and add 1. 4. Possible rule: Double the number of the stage and add 1.

Station 2

Students work together to analyze a pattern based on the cost of a gym membership. They work together to extend a table of values and predict the cost of joining the gym for various numbers of months. The emphasis is on recognizing and describing the underlying numerical pattern.

Answers

1. $92, $104; 2. Possible explanation: Extend the table. For each additional month, the cost increases by $12; 3. 10 months. 4. Possible explanation: Extend the table until the cost is $140 and read the corresponding number of months. 5. Possible description: For each additional month, the cost increases by $12. (Although students are not required to use variables, they may notice that the cost for n months is $20 + 12n$.)

Station 3

Students work together to analyze and extend a pattern that is provided in the form of a table. The table shows the number of baseball cards in a collection from one week to the next. Students are asked to predict the number of cards in future weeks and to describe the pattern.

Answers

1. 87; 2. Possible explanation: Extend the table to the ninth week by adding 8 cards per week. 3. 14 weeks; 4. Extend the table until the number of cards is greater than 120 and count the number of weeks. 5. Possible description: For each additional week, the number of cards increases by 8. (Although students are not required to use variables, they may notice that the number of cards on week n is $15 + 8n$.)

Station 4

Students build, extend, and describe a pattern based on tables and chairs. To do so, they may use physical objects to model the pattern. Students work together to determine how many chairs are needed for a given number of tables, and then generalize their work by describing a rule for the pattern.

Answers

1. 10 chairs; 12 chairs; 2. 26 chairs; 3. Possible explanation: Double the number of tables and add 2. 4. Possible rule: Multiply the number of tables by 2 and then add 2.

Materials List/Setup

Station 1	box of toothpicks
Station 2	none
Station 3	none
Station 4	small square tiles and small round tiles

Discussion Guide

To support students in reflecting on the activities, and to gather formative information about student learning, use the following prompts to facilitate a class discussion to "debrief" the station activities.

Prompts/Questions

1. What are some different tools, objects, or drawings that you can use to help you analyze a pattern?

2. What are some strategies you can use to help extend a number pattern?

3. How can you use a calculator to check your work when you extend a number pattern?

4. Can all the patterns in these activities be extended indefinitely? Why or why not?

Think, Pair, Share

Have students jot down their own responses to questions, discuss their responses with a partner (who was not in their station group), and then discuss as a whole class.

Suggested Appropriate Responses

1. Hands-on manipulatives, drawings of patterns, tables, etc.

2. Look for a pattern in the sequence of numbers. For example, a fixed number may be added each time you go from one number to the next.

3. Use the calculator to check that the repeated addition was performed correctly.

4. Yes. Each pattern is based on adding a fixed number at each stage of the pattern. That process can be continued forever.

Possible Misunderstandings/Mistakes

*   Incorrectly determining the value by which a numerical pattern increases or decreases from one term to the next

*   Assuming that all patterns may be described multiplicatively (i.e., in the form $y = kx$)

*   Incorrectly reproducing a given pattern with manipulatives or with a drawing

Ratios and Proportional Relationships
Set 2: Patterns and Relationships

Station 1

You will find some toothpicks at this station. Use them to help you with this activity.

Here are several stages of a pattern made from toothpicks.

Stage 1 **Stage 2** **Stage 3**

Work with other students to build the pattern from toothpicks.

Then work together to answer the following questions. When everyone agrees on an answer, write it in the space provided.

1. How many toothpicks do you need to make Stage 4 and Stage 5? _____

2. Predict the number of toothpicks you would need to make Stage 10. _____

3. Explain how you made this prediction. _____

   _____

   _____

   _____

4. Describe a rule for the pattern. That is, if you are given the stage of the pattern, describe a rule for finding the number of toothpicks needed.

   _____

   _____

   _____

Ratios and Proportional Relationships
Set 2: Patterns and Relationships

Station 2

At this station, you will explore a pattern based on prices.

A gym posts this table at the front desk. It shows the cost of joining the gym for different numbers of months.

Number of months	1	2	3	4	5
Cost of membership	$32	$44	$56	$68	$80

Work together to answer the following questions. When everyone agrees on an answer, write it in the space provided.

1. Predict the cost of joining the gym for 6 months and for 7 months. _____

2. Explain how you made these predictions. _____

_____

_____

_____

3. Janelle has $140 to spend on a membership to this gym. For how many months can she join?

_____

4. Explain how you found the answer to Question 3. _____

_____

_____

_____

5. Describe the pattern in the table. _____

_____

_____

_____

Ratios and Proportional Relationships
Set 2: Patterns and Relationships

Station 3

At this station, you will explore a pattern based on a table of data.

Diego collects baseball cards. Each week, he buys some new cards. He starts with 23 cards in his collection the first week. The table below shows the number of cards in Diego's collection each week.

Week	1	2	3	4	5
Number of cards	23	31	39	47	55

Work together to answer the following questions. When everyone agrees on an answer, write it in the space provided.

1. Predict the number of cards in Diego's collection on the 9th week. _____

2. Explain how you made this prediction. _____
   _____
   _____
   _____

3. How many weeks will it take Diego to have more than 120 cards in his collection?
   _____

4. Explain how you found the answer to Question 3. _____
   _____
   _____
   _____

5. Describe the pattern in the table. _____
   _____
   _____
   _____

Ratios and Proportional Relationships
Set 2: Patterns and Relationships

Station 4

At this station, you will find some tiles that you may use to help you analyze a pattern.

A restaurant has square tables that can be pushed together in a row to seat large groups. The figure below shows the number of chairs (circles) that are needed for various numbers of tables (squares).

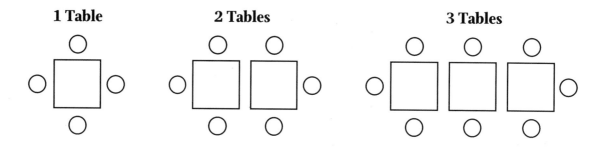

Work together to answer the following questions. When everyone agrees on an answer, write it in the space provided.

1. How many chairs are needed for 4 tables? for 5 tables? _____

2. Predict the number of chairs that are needed for 12 tables. _____

3. Explain how you made this prediction. _____

    _____

    _____

    _____

4. Describe a rule for the pattern. That is, if you are given the number of tables, describe a rule for finding the number of chairs.

    _____

    _____

    _____

Ratios and Proportional Relationships

Set 3: Proportional Relationships

Goal: To provide opportunities for students to develop concepts and skills related to proportional relationships

Common Core Standards

Ratios and Proportional Relationships

Understand ratio concepts and use ratio reasoning to solve problems.

6.RP.1. Understand the concept of a ratio and use ratio language to describe a ratio relationship between two quantities.

6.RP.2. Understand the concept of a unit rate a/b associated with a ratio $a{:}b$ with $b \neq 0$, and use rate language in the context of a ratio relationship.

6.RP.3. Use ratio and rate reasoning to solve real-world and mathematical problems, e.g., by reasoning about tables of equivalent ratios, tape diagrams, double number line diagrams, or equations.

    a. Make tables of equivalent ratios relating quantities with whole-number measurements, find missing values in the tables, and plot the pairs of values on the coordinate plane. Use tables to compare ratios.

Student Activities Overview and Answer Key

Station 1

Students are given a set of cards with ratios written on them. Students work together to pair up the cards so that the ratios shown on each pair of cards are equal. After forming the pairs of cards, students explain the strategies they used to solve the problem.

Answers

The cards should be paired as follows: $^2/_3 = ^6/_9$; $^4/_8 = ^8/_{16}$; $^5/_{15} = \ ^2/_6$; $^{12}/_{15} = ^8/_{10}$; $^3/_4 = ^6/_8$

Possible strategies: Look for pairs of cards that represent common ratios, such as 1:2. Reduce fractions to lowest terms and look for equivalent pairs. Pair up cards and check to see if the ratios are equal using cross products.

Station 2

Students work together to explore a real-world situation that involves a proportional relationship. First, students look for patterns in a table of data and complete the table. Then they make predictions about the data in the table and write an equation in the form of $y = kx$ that describes the relationship.

Answers

1. See below for complete table. 2. Possible explanation: The number of ounces of meat divided by the number of feedings is always 2.4. 3. 2.4 ounces; 4. 48 ounces; 5. $y = 2.4x$; 6. Since x is the number of feedings, you must multiply it by 2.4 to get the number of ounces of meat. So the relationship is $y = 2.4x$.

Number of feedings	3	4	6	9	12
Ounces of meat	7.2	9.6	14.4	21.6	28.8

Station 3

Students learn a step-by-step method for using graph paper and a straightedge to check whether two ratios are proportional. Then students work together to use this method to test several pairs of ratios. They also use the method to find a ratio that is proportional to a given ratio.

Answers

1. yes; 2. yes; 3. no; 4. yes; 5. no; 6. yes; 7. Possible ratio: $1/4$; Draw a 3-by-12 rectangle and its diagonal. Then look for a smaller or larger rectangle that could have the same line as a diagonal.

Station 4

Students work together to use the numbers 2, 3, 4, and 6 to write as many different proportions as they can. To help them, the numbers are provided on cards that can be moved around and paired in different ways. Students explain the strategies they used to find the proportions.

Answers

Possible proportions: $2/3 = 4/6$; $2/4 = 3/6$; $3/2 = 6/4$; $4/2 = 6/3$

Possible strategies: Place the numbers in any positions and use cross products to check whether the ratios form a proportion. Try to pair the numbers to make common ratios, such as 1:2.

Materials List/Setup

Station 1	set of 12 index cards with the following ratios written on them:
	$2/3$, $6/9$, $8/10$, $12/15$, $3/4$, $6/8$, $4/8$, $8/16$, $2/6$, $5/15$
Station 2	calculator
Station 3	graph paper; straightedge
Station 4	set of 4 index cards with the following numbers written on them:
	2, 3, 4, 6

Discussion Guide

To support students in reflecting on the activities, and to gather formative information about student learning, use the following prompts to facilitate a class discussion to "debrief" the station activities.

Prompts/Questions

1. Is the ratio $^2/_3$ the same as the ratio $^3/_2$? Why or why not?

2. What does it mean when we say that two ratios form a proportion?

3. How can you check that two ratios form a proportion?

4. If you are given a ratio, how can you form a different ratio that is proportional to the given one?

Think, Pair, Share

Have students jot down their own responses to questions, discuss their responses with a partner (who was not in their station group), and then discuss as a whole class.

Suggested Appropriate Responses

1. No. The ratio $^2/_3$ is less than 1, while the ratio $^3/_2$ is greater than 1.

2. The ratios are equivalent. They name the same fraction.

3. Check to see if the cross products are equal.

4. Form a new ratio by multiplying the numerator and denominator of the given ratio by the same number.

Possible Misunderstandings/Mistakes

- Assuming that a ratio and its reciprocal are equivalent ratios

- Adding a number to the numerator and denominator of a given ratio to form a new ratio that is proportional to the given one

- Failing to recognize that whole numbers may be written as ratios by using a denominator of 1

Ratios and Proportional Relationships
Set 3: Proportional Relationships

Station 1

At this station, you will find a set of cards with the following ratios written on them:

$$\frac{2}{3} \qquad \frac{4}{8} \qquad \frac{5}{15} \qquad \frac{12}{15} \qquad \frac{3}{4} \qquad \frac{6}{8} \qquad \frac{6}{9} \qquad \frac{8}{16} \qquad \frac{2}{6} \qquad \frac{8}{10}$$

Work with other students to put the cards in pairs. The ratios in each pair should be proportional.

Work together to check that the ratios are proportional. When everyone agrees on your answers, write the pairs of equal ratios below.

_____

_____

_____

_____

_____

Explain at least three strategies that you used to solve this problem.

_____

_____

_____

_____

_____

_____

_____

_____

_____

Station Activities for Common Core Mathematics, Grade 6

Ratios and Proportional Relationships
Set 3: Proportional Relationships

Station 2

At this station, you will work with other students to explore a real-world situation.

A zookeeper is taking care of a baby tiger. The table below shows the number of feedings and the number of ounces of meat that are needed.

Number of feedings	3	4	6	9	12
Ounces of meat	7.2	9.6	14.4		

Work with other students to look for patterns in the table. You may use a calculator to help you.

1. Complete the table.

2. Explain how you completed the table. _____

_____

_____

_____

3. How many ounces of meat are needed for each feeding? _____

4. How many ounces of meat are needed for 20 feedings? _____

5. Write an equation that shows the number of ounces of meat y that are needed for x feedings.

_____

6. Explain how you came up with the equation.

_____

_____

_____

Station Activities for Common Core Mathematics, Grade 6

Ratios and Proportional Relationships

Set 3: Proportional Relationships

Station 3

You will find graph paper and a straightedge at this station. You will use these tools to check whether ratios are proportional.

To check whether $\frac{2}{3}$ and $\frac{4}{6}$ are proportional, first draw a

2-by-3 rectangle on the graph paper.

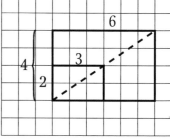

Now draw a 4-by-6 rectangle. The two rectangles should share a corner as shown.

$$\frac{2}{3} = \frac{4}{6}$$

Use the straightedge to draw the diagonals of the rectangles. If you can draw the diagonals of both rectangles with one straight line, then the ratios are proportional.

Work with other students to use this method to check whether the ratios in each pair are proportional. Write "yes" or "no" for each pair.

1. $\frac{1}{2}$ and $\frac{4}{8}$ _____

2. $\frac{6}{8}$ and $\frac{9}{12}$ _____

3. $\frac{5}{8}$ and $\frac{10}{14}$ _____

4. $\frac{2}{5}$ and $\frac{4}{10}$ _____

5. $\frac{6}{10}$ and $\frac{8}{12}$ _____

6. $\frac{4}{6}$ and $\frac{6}{9}$ _____

7. Use this method to find a ratio that is proportional to $\frac{3}{12}$. Write the ratio below and explain your work.

_____

_____

_____

_____

Ratios and Proportional Relationships
Set 3: Proportional Relationships

Station 4

At this station, you will work with other students to form ratios that are proportional.

You will find cards at this station with the numbers 2, 3, 4, and 6 on them.

Work with other students to write the four numbers in the boxes below so that the two ratios are proportional.

$$\frac{\square}{\square} = \frac{\square}{\square}$$

Use the cards to help you move the numbers around and form other ratios that are proportional. Work together to find as many different proportions as you can.

Record your proportions below.

_____

_____

_____

Explain the strategies you used to find the proportions.

_____

_____

_____

_____

The Number System

Goal: To provide opportunities for students to develop concepts and skills related to absolute values

Common Core Standards

The Number System

Apply and extend previous understandings of numbers to the system of rational numbers.

6.NS.5. Understand that positive and negative numbers are used together to describe quantities having opposite directions or values (e.g., temperature above/below zero, elevation above/below sea level, credits/debits, positive/negative electric charge); use positive and negative numbers to represent quantities in real-world contexts, explaining the meaning of 0 in each situation.

6.NS.6. Understand a rational number as a point on the number line.

c. Find and position integers and other rational numbers on a horizontal or vertical number line diagram; find and position pairs of integers and other rational numbers on a coordinate plane.

6.NS.7. Understand ordering and absolute value of rational numbers.

c. Understand the absolute value of a rational number as its distance from 0 on the number line; interpret absolute value as magnitude for a positive or negative quantity in a real-world situation.

Student Activities Overview and Answer Key

Station 1

Students work together to use a number line to plot numbers that have a given absolute value. Students reflect on their work to recognize that there are two points on a number line with a given absolute value (for positive absolute values) and one point (0) with an absolute value of zero.

Answers

1. 1 and –1; 2. 3 and –3; 3. 4.5 and –4.5; 4. 0; 5. 2 and –2

Possible responses: For a given positive absolute value, two points have the given absolute value. The only point with an absolute value of 0 is 0. No points have a negative absolute value.

Station 2

Students use a penny and a number cube to generate positive and negative two-digit numbers. They work together to find the absolute value of each of the numbers.

Answers

Answers will depend upon numbers that are rolled.

Possible explanation: If the number is positive, the absolute value of the number is the same as the number. If the number is negative, the absolute value of the number is the same as the number without its sign.

Station 3

Students are given a set of cards with positive and negative numbers written on them. Students work together to arrange the cards so that they are in order from the number with the least absolute value to the number with the greatest absolute value.

Answers

0, −1, −1.5, 2, −3, 4.1, 8, −12

Possible strategies: First write the absolute value of each number. Then list the numbers so that the absolute values are listed in increasing order.

Station 4

Students are given a set of cards with numbers or expressions written on them. Students work together to form pairs of cards so that the numbers in each pair have the same absolute value. Then students reflect on the strategies they used to solve the problem.

Answers

6 + 1 and −7; 5 and −5; 8 ÷ 4 and −2; 3 and 9 − 12; 2 + 8 and 2 − 12

Possible strategies: First simplify the expressions that involve operations. Look for pairs of numbers that are opposites.

Materials List/Setup

Station 1 none

Station 2 a penny; number cube (numbered 1–6)

Station 3 8 index cards with the following numbers written on them:

0, −1, −1.5, 2, −3, 4.1, 8, −12

Station 4 10 index cards with the following numbers or expressions written on them:

6 + 1, −7, 5, −5, 8 ÷ 4, −2, 3, 9 − 12, 2 + 8, 2 − 12

Discussion Guide

To support students in reflecting on the activities and to gather some formative information about student learning, use the following prompts to facilitate a class discussion to "debrief" the station activities.

Prompts/Questions

1. What is the absolute value of a number?

2. How do you find the absolute value of a number?

3. Can the absolute value of a number ever be negative? Why or why not?

4. Can two different numbers ever have the same absolute value? Explain.

Think, Pair, Share

Have students jot down their own responses to questions, then discuss with a partner (who was not in their station group), and then discuss as a whole class.

Suggested Appropriate Responses

1. the distance of the number from 0 on a number line

2. If the number is positive, the absolute value of the number is the same as the number. If the number is negative, the absolute value of the number is the same as the number without its sign. The absolute value of 0 is 0.

3. No. The absolute value represents a distance, so it cannot be negative.

4. Yes. A number and its opposite, such as 5 and –5, have the same absolute value.

Possible Misunderstandings/Mistakes

- Forgetting that the absolute value of a number cannot be negative

- Incorrectly finding the absolute value of expressions (e.g., $|2 - 12| \neq |2| - |12|$)

- Incorrectly ordering the absolute value of two numbers (e.g., stating that $|-5|$ is less than $|4|$ because –5 is less than 4)

The Number System
Set 1: Integers and Absolute Value

Station 1

In this activity, you will work with other students to plot points on a number line. Recall that the absolute value of a number is its distance from zero on a number line.

For each of the following, plot all the points (if any) that satisfy the statement.

1. The absolute value is 1.

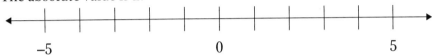

2. The absolute value is 3.

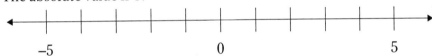

3. The absolute value is 4.5.

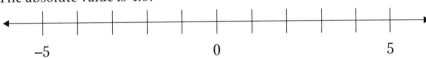

4. The absolute value is 0.

5. The absolute value is –2.

Work with other students to list at least three things you notice from your work. _____

_____

_____

_____

The Number System
Set 1: Integers and Absolute Value

Station 2

You will find a penny and a number cube at this station. You will use these to create positive and negative numbers.

Flip the penny. If it lands heads up, write "+" on the line below. If it lands tails up, write "−" on the line. Then roll the number cube two times. Write the numbers in the two boxes below.

_____ ☐ ☐

Find the absolute value of the number. _____

Repeat the process four more times.

_____ ☐ ☐   _____ ☐ ☐

Absolute value: _____   Absolute value: _____

_____ ☐ ☐   _____ ☐ ☐

Absolute value: _____   Absolute value: _____

Explain how you found the absolute values of the numbers.

_____

_____

_____

The Number System
Set 1: Integers and Absolute Value

Station 3

You will find a set of cards at this station. The cards have the following numbers written on them.

$$-1.5 \qquad 4.1 \qquad 0 \qquad -12 \qquad 8 \qquad -1 \qquad -3 \qquad 2$$

Work with other students to arrange the cards so that they are in order from the number with the least absolute value to the number with the greatest absolute value.

Write your answer below.

_____

Work together to check that the numbers are in the correct order.

Describe the strategies you used to solve this problem.

_____

_____

_____

_____

_____

The Number System
Set 1: Integers and Absolute Value

Station 4

You will be given a set of index cards with the following numbers and expressions on them:

$6 + 1$ 5 $9 - 12$ -2 -7

-5 $2 - 12$ 3 $8 \div 4$ $2 + 8$

Work with other students to form pairs of cards. The two numbers or expression in each pair should have the same absolute value.

Work together to check that the numbers or expressions in each pair have the same absolute value. Write the pairs below.

_____

_____

_____

_____

_____

Describe any strategies you used to solve this problem.

The Number System

Goal: To provide opportunities for students to develop concepts and skills related to comparing and ordering numbers

Common Core Standards

The Number System

Apply and extend previous understandings of numbers to the system of rational numbers.

6.NS.5. Understand that positive and negative numbers are used together to describe quantities having opposite directions or values (e.g., temperature above/below zero, elevation above/below sea level, credits/debits, positive/negative electric charge); use positive and negative numbers to represent quantities in real-world contexts, explaining the meaning of 0 in each situation.

6.NS.6. Understand a rational number as a point on the number line.

c. Find and position integers and other rational numbers on a horizontal or vertical number line diagram; find and position pairs of integers and other rational numbers on a coordinate plane.

6.NS.7. Understand ordering and absolute value of rational numbers.

c. Understand the absolute value of a rational number as its distance from 0 on the number line; interpret absolute value as magnitude for a positive or negative quantity in a real-world situation.

Student Activities Overview and Answer Key

Station 1

Each student holds a sheet of paper that shows a decimal between 0 and 1. Students work together to line up so that the decimals are in order from least to greatest. When the students are in a line, they work together to check that the numbers have been ordered correctly.

Answers

$0.02, 0.05, 0.1, 0.2, 0.\overline{2}, 0.5, 0.51, 0.\overline{5}$

First write out several digits of the repeating decimals (e.g., $0.2 = 0.22\overline{2}\ldots$). Order the decimals by comparing the tenths digits of the numbers. If the tenths digits are the same, compare the hundredths digits.

Station 2

Students use a penny and a number cube to create a set of five decimals that may be positive or negative. Then they work together to put the five decimals in order from least to greatest.

Answers

Answers will vary depending upon the numbers rolled.

Possible strategies: Any negative decimal is less than any positive decimal. If two decimals have the same sign, first compare the units digits. If these are the same, compare the tenths digits.

Station 3

Students are given eight positive and negative numbers in the form of fractions, decimals, and percents. The numbers are written on index cards and students work together to arrange the cards so the numbers go from least to greatest. Then they work together to plot the numbers on a number line.

Answers

$-4.5, -2\frac{1}{2}, -1, -0.5, 50\%, 1.0, 3\frac{1}{2}, 4$

Possible strategies: Convert all the numbers to decimals; first order the negative numbers by placing them to the left of 0 on the number line, then order the positive numbers by placing them to the right of 0 on the number line.

Station 4

Students are given ten numbers in the form of fractions and decimals. The numbers are written on index cards and students work together to find pairs of cards that represent equal numbers. Then students write equality statements based on the pairs of matching cards.

Answers

$1\frac{1}{3} = 1.\overline{3}; \frac{2}{3} = 0.\overline{6}; \frac{4}{5} = 0.8; 2\frac{3}{10} = 2.3; 1\frac{1}{5} = 1.2$

Possible strategies: Find familiar benchmarks (e.g., $\frac{2}{3} = 0.\overline{6}$); convert the fractions to decimals; convert the decimals to fractions.

Materials List/Setup

Station 1   8 sheets of paper with the following numbers written on them (large enough to be seen from a distance):
$0.02, 0.05, 0.1, 0.2, 0.\overline{2}, 0.5, 0.51, 0.\overline{5}$
The sheets should be shuffled and placed face down on a desk or table at the station.

Station 2   a penny; number cube (numbered 1–6)

Station 3   8 index cards with the following numbers written on the cards:

$-4.5, -2\frac{1}{2}, -1, -0.5, 50\%, 1.0, 3\frac{1}{2}, 4$

Station 4   10 index cards with the following numbers written on the cards:

$1\frac{1}{3}, 1.\overline{3}, \frac{2}{3}, 0.\overline{6}, 2\frac{3}{10}, 2.3, \frac{4}{5}, 0.8, 1\frac{1}{5}, 1.2$

Discussion Guide

To support students in reflecting on the activities and to gather some formative information about student learning, use the following prompts to facilitate a class discussion to "debrief" the station activities.

Prompts/Questions

1. What are some different strategies that you can use to help you compare various types of numbers?

2. How do you compare two decimals?

3. What can you say about two numbers if one is positive and the other is negative?

4. How can you compare a fraction and a decimal to decide which is greater?

Think, Pair, Share

Have students jot down their own responses to questions, then discuss with a partner (who was not in their station group), and then discuss as a whole class.

Suggested Appropriate Responses

1. Plot the points on a number line; convert different types of numbers (e.g., fractions, decimals, percents, etc.) to the same form so they can be compared more easily.

2. Compare the digits from left to right. For example, for decimals between 0 and 1, first compare tenths digits. If these are the same, compare hundredths digits. If these are the same, compare thousandths digits, and so on.

3. The positive number is greater than the negative number.

4. Convert the fraction to a decimal. Then compare the two decimals. Alternatively, compare the fraction and decimal to familiar benchmark values.

Possible Misunderstandings/Mistakes

- Assuming longer decimals represent greater numbers (e.g., that 0.55 is greater than 0.6)

- Incorrectly converting fractions to decimals

- Comparing signed numbers without taking the signs into account (e.g., stating that −7.5 is greater than 7.4)

The Number System
Set 2: Comparing and Ordering Rational Numbers

Station 1

You will find some sheets of paper at this station. Each student in your group should choose one of the sheets of paper. (It's okay if there are some sheets of paper left over.)

1. Each student should hold his or her sheet of paper so everyone can see the number on it.

2. Work as a group to line up so that the numbers are in order from least to greatest.

3. When everyone is in a line, work together to check that the numbers are in order.

4. Write the numbers in order on the line below.

_____

Write at least three strategies you could use to put the numbers in order.

_____

_____

_____

_____

_____

_____

_____

The Number System

Set 2: Comparing and Ordering Rational Numbers

Station 2

You will find a penny and a number cube at this station. You will use these to create positive and negative decimals.

Flip the penny. If it lands heads up, write "+" on the line. If it lands tails up, write "−" on the line. Then roll the number cube two times. Write the numbers in the two boxes below.

Repeat the process to create four more decimals.

Work as a group to put your five decimals in order from least to greatest. Write the decimals in order below.

Explain how you put the numbers in order.

Station Activities for Common Core Mathematics, Grade 6   © 2011 Walch Education

The Number System
Set 2: Comparing and Ordering Rational Numbers

Station 3

At this station, you will find a set of cards with the following numbers written on them:

$$1.0 \qquad -0.5 \qquad -2\frac{1}{2} \qquad 3\frac{1}{2} \qquad -1 \qquad 4 \qquad 50\% \qquad -\frac{4}{5}$$

Work with other students to put the cards in order so the numbers go from least to greatest. Then plot and label the numbers on the number line. Work as a group to make sure the numbers are all plotted correctly.

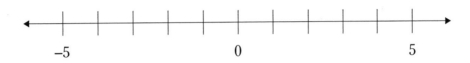

Write at least three strategies you could use to help you decide how to put the numbers in order.

_____

_____

_____

_____

_____

_____

_____

_____

NAME:

The Number System
Set 2: Comparing and Ordering Rational Numbers

Station 4

At this station, you will find a set of cards with the following numbers written on them.

$\frac{4}{5}$ $2\frac{3}{10}$ $0.\overline{6}$ $1\frac{1}{3}$ $1\frac{1}{5}$ 2.3 1.2 0.8 $1.\overline{3}$ $\frac{2}{3}$

Work with other students to find pairs of cards that show the same number. When you have paired up the cards, work as a group to check that the numbers in each pair are equal.

Write five statements that use an equal sign (=) to list the pairs of equal numbers.

_____

_____

_____

_____

_____

Write at least three strategies you could use to help you decide which numbers are equal.

_____

_____

_____

_____

_____

_____

_____

_____

The Number System

Set 3: Factors, Multiples, and Prime Factorization

Goal: To provide opportunities for students to develop concepts and skills related to factors, multiples, and prime factorizations

Common Core Standards

The Number System

Compute fluently with multi-digit numbers and find common factors and multiples.

6.NS.4. Find the greatest common factor of two whole numbers less than or equal to 100 and the least common multiple of two whole numbers less than or equal to 12.

Student Activities Overview and Answer Key

Station 1

Students use a number cube to generate two-digit numbers. Then they work together to decide if each of the numbers is prime or composite. If the number is composite, they find the prime factorization. Finally, students describe the strategies they used to identify the numbers as prime or composite.

Answers

Answers will depend on the numbers that are rolled.

Possible strategies: Any two-digit even number is composite; any number ending in 5 is divisible by 5, so it is composite; recognize familiar primes, such as 11 and 13.

Station 2

Students are given 12 red tiles and 18 blue tiles. They are asked to arrange the tiles in rows so that each row contains the same number of tiles and so that each row contains only red tiles or only blue tiles. The rows must also be as long as possible. When students have arranged the tiles following these rules, they reflect on how their arrangement is related to the greatest common factor of 12 and 18.

Answers

Each row has 6 tiles. There are 2 rows with 6 red tiles in each row and 3 rows with 6 blue tiles in each row. The number of tiles in each row (6) is the greatest common factor of 12 and 18.

Station 3

Students use a highlighter to highlight all the numbers less than or equal to 100 that are multiples of 8. Then they use a different color to highlight all the multiples of 12. Students work together to check that the multiples are highlighted correctly. Then they look for common multiples (numbers highlighted in both colors) and identify the least common multiple.

Answers

Multiples of 8: 8, 16, 24, 32, 40, 48, 56, 64, 72, 80, 88, 96

Multiples of 12: 12, 24, 36, 48, 60, 72, 84, 96

Numbers highlighted in both colors (common multiples): 24, 48, 72, 96

The least common multiple of 8 and 12 is the smallest of the common multiples, 24.

Station 4

Students are given a set of cards with numbers on them. They choose two cards at random and work together to find the greatest common factor and least common multiple of the two numbers chosen. Students replace the cards, shuffle them, and repeat the process two more times. Then they reflect on the strategies they used.

Answers

Answers will depend upon the cards that are chosen.

Possible strategies: To find the greatest common factor, list all the factors of each number and choose the greatest factor that appears in both lists. To find the least common multiple, list several multiples of each number and choose the smallest multiple that appears in both lists.

Materials List/Setup

Station 1	number cube (numbered 1–6)
Station 2	12 small red tiles; 18 small blue tiles
Station 3	two highlighters (different colors)
Station 4	8 index cards with the following numbers written on them:
	4, 6, 8, 10, 12, 15, 16, 18

Discussion Guide

To support students in reflecting on the activities and to gather some formative information about student learning, use the following prompts to facilitate a class discussion to "debrief" the station activities.

Prompts/Questions

1. What are some ways to decide if a number is prime or composite?

2. Is it ever possible for an even number to be prime? Why or why not?

3. Suppose you had 98 red tiles and 66 blue tiles, and you wanted to arrange all of them in rows so that each row contained only red tiles or only blue tiles, and so that the rows were as long as possible. How could you find the number of tiles in each row without actually arranging all the tiles?

4. Can the least common multiple of two numbers ever be one of the given numbers? Explain.

Think, Pair, Share

Have students jot down their own responses to questions, then discuss with a partner (who was not in their station group), and then discuss as a whole class.

Suggested Appropriate Responses

1. Possible methods: Aside from 2, any even number is composite. A prime number cannot have a units digit of 5. Check to see if the number is divisible by any of the prime numbers that are smaller than it.

2. Yes; the number 2 is prime. Any other even number is not prime because it is divisible by 2.

3. Find the greatest common factor of 98 and 66.

4. Yes. For example, the least common multiple of 2 and 10 is 10.

Possible Misunderstandings/Mistakes

- Assuming that odd numbers are prime (e.g., stating that 39 is prime)

- Multiplying two numbers to find the least common multiple (e.g., stating that the least common multiple of 6 and 8 is 48)

- Forgetting that the greatest common factor of two numbers may be one of the numbers

The Number System

Set 3: Factors, Multiples, and Prime Factorization

Station 1

You will find a number cube at this station. You will use the number cube to create some two-digit numbers.

Roll the number cube two times. Write the numbers in the boxes below.

Work with other students to decide if this two-digit number is prime or composite. Write the answer on the line. _____

If the number is composite, write the prime factorization of the number on the line below.

_____

Repeat the process two more times.

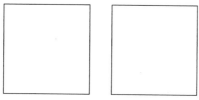

Prime or composite? _____

If composite, prime factorization: _____

Prime or composite? _____

If composite, prime factorization: _____

Write at least three strategies you used to help you decide whether each number was prime or composite.

_____

_____

_____

_____

The Number System
Set 3: Factors, Multiples, and Prime Factorization

Station 2

At this station, you will find 12 red tiles and 18 blue tiles.

Work with other students to arrange the tiles in rows. You must follow these rules:

- Use all the tiles.
- A row can contain only blue tiles or only red tiles.
- Each row must contain the same number of tiles.
- The rows should be as long as possible.

Sketch your arrangement of tiles in the space below.

Work together to check that your arrangement matches all the rules.

How many tiles are in each row? _____

Explain how the number of tiles in each row is related to the greatest common factor.

_____

_____

_____

_____

The Number System
Set 3: Factors, Multiples, and Prime Factorization

Station 3

You will find two highlighters at this station. Use one highlighter to highlight all the numbers in the grid that are multiples of 8. Use the other highlighter to highlight all the numbers in the grid that are multiples of 12.

1	2	3	4	5	6	7	8	9	10
11	12	13	14	15	16	17	18	19	20
21	22	23	24	25	26	27	28	29	30
31	32	33	34	35	36	37	38	39	40
41	42	43	44	45	46	47	48	49	50
51	52	53	54	55	56	57	58	59	60
61	62	63	64	65	66	67	68	69	70
71	72	73	74	75	76	77	78	79	80
81	82	83	84	85	86	87	88	89	90
91	92	93	94	95	96	97	98	99	100

Work together to check that you have highlighted the multiples correctly.

Which numbers are highlighted in both colors? _____

What can you say about these numbers? _____

Explain how to use your work to find the least common multiple of 8 and 12.

_____

_____

_____

The Number System

Set 3: Factors, Multiples, and Prime Factorization

Station 4

You will be given a set of index cards. Shuffle the cards and place them face down.

Choose two cards without looking. Turn the cards over. Write the two numbers below.

_____   _____

Work with other students to find the greatest common factor and least common multiple of the numbers.

Greatest common factor: _____

Least common multiple: _____

Put the cards back. Shuffle the cards. Then repeat the above process two more times.

Two numbers: _____   _____

Greatest common factor: _____

Least common multiple: _____

Two numbers: _____   _____

Greatest common factor: _____

Least common multiple: _____

Explain the strategies you used to find the greatest common factors.

_____

_____

_____

Explain the strategies you used to find the least common multiples.

_____

_____

_____

The Number System

Set 4: Multiplying and Dividing Fractions

Goal: To provide students with opportunities to develop concepts and skills related to multiplication and division

Common Core Standards

The Number System

Apply and extend previous understandings of multiplication and division to divide fractions by fractions.

> **6.NS.1.** Interpret and compute quotients of fractions, and solve word problems involving division of fractions by fractions, e.g., by using visual fraction models and equations to represent the problem.

Student Activities Overview and Answer Key

Station 1

Students model the product of two fractions using 10-by-10 grids. To do so, students model one fraction in the product by shading columns and the other by shading rows. The overlapping region represents the product of the fractions. After working through a step-by-step example of the method, students work together to use the method to find additional products.

Answers

$^3/_{10} \times ^1/_2 = ^3/_{20}$

1. $^1/_{50}$; 2. $^1/_5$; 3. $^1/_{25}$; 4. $^1/_{100}$

Station 2

Students use rectangular strips of paper to help them divide fractions. They first fold the strips into sixteen equal sections. Given a division problem, students work together to model the first fraction (the dividend) by coloring sections of the strip yellow. Within the yellow section of the strip, students color sections blue to model the divisor. The number of times the blue section would fit inside the yellow section models the answer to the division problem.

Answers

$^3/_8 \div ^1/_{16} = 6$

1. 4; 2. 12; 3. 4; 4. 2; 5. 2

Station 3

Students are given two sets of cards with fractions written on them. They choose one card from each set and work together to find the quotient of the fractions. At the end of the activity, students discuss the strategies they used to divide the fractions.

Answers

Answers will depend upon the cards that are chosen.

Possible strategies: Divide the fractions by changing the operation to multiplication and changing the divisor to its reciprocal. Then multiply. Simplify the answer.

Station 4

Students are given cards with fractions or mixed numbers written on them. They use the cards and a penny to create various expressions involving multiplication and division of fractions and mixed numbers. Students work together to perform the required operation and simplify the resulting fraction.

Answers

Answers will depend upon the numbers that are chosen.

Materials List/Setup

Station 1	red and blue pencils
Station 2	yellow and blue highlighters rectangular strips of paper, about 1 inch wide and 1 foot long (These may be cut from sheets of paper, or you can use strips of adding-machine tape.)
Station 3	10 index cards, prepared as follows:

- set of 5 cards, labeled "A" on the back with the following fractions written on the front: $3/4$, $4/5$, $5/6$, $15/16$, $5/12$
- set of 5 cards, labeled "B" on the back with the following fractions written on the front: $1/6$, $1/3$, $5/8$, $1/4$, $2/5$
- Place the cards face-down at the station, so that only the backs (A or B) are visible.

Station 4	penny 8 index cards with the following fractions or mixed numbers written on them: $1/5$, $3/3$, $3/4$, $5/8$, $4/5$, $1\frac{1}{2}$, $1\frac{2}{3}$, $2\frac{3}{4}$

Discussion Guide

To support students in reflecting on the activities and to gather some formative information about student learning, use the following prompts to facilitate a class discussion to "debrief" the station activities.

Prompts/Questions

1. How do you multiply two fractions?

2. How do you multiply two fractions if one of them is a mixed number?

3. How do you divide two fractions?

4. What are some different models, drawings, or tools that you can use to help you multiply and divide fractions?

Think, Pair, Share

Have students jot down their own responses to questions, then discuss with a partner (who was not in their station group), then discuss as a whole class.

Suggested Appropriate Responses

1. Multiply the numerators and multiply the denominators.

2. First change the mixed number to an improper fraction. Then multiply as usual.

3. Change the operation to multiplication and change the divisor to its reciprocal. Then multiply the fractions as usual and simplify the result.

4. 10-by-10 grids, strips of paper, calculators, etc.

Possible Misunderstandings/Mistakes

- Forgetting to take the reciprocal of the divisor when converting a division problem to a multiplication problem

- Not recognizing that whole numbers may be written as fractions in products or quotients by writing them with a denominator of 1

- Incorrectly converting a mixed number to an improper fraction or vice versa

The Number System
Set 4: Multiplying and Dividing Fractions

Station 1

Each 10-by-10 grid represents 1. Each column of the grid represents $\frac{1}{10}$. Each row also represents $\frac{1}{10}$.

You can use the grids to multiply fractions. To multiply $\frac{3}{10} \times \frac{1}{2}$, first use a red pencil to shade columns that show $\frac{3}{10}$ of the grid.

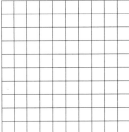

Next, use a blue pencil to shade rows that show $\frac{1}{2}$ of the grid.

Work with other students to figure out what fraction of the entire grid is shaded both red and blue. This is the product. Write it below.

_____

Work with other students and use this method to find each product.

1. $\frac{1}{5} \times \frac{1}{10} =$ _____

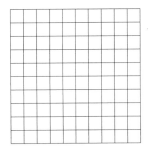

2. $\frac{1}{2} \times \frac{2}{5} =$ _____

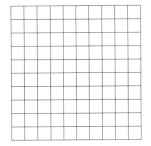

3. $\frac{1}{10} \times \frac{2}{5} =$ _____

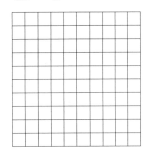

4. $\frac{1}{10} \times \frac{1}{10} =$ _____

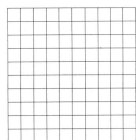

The Number System
Set 4: Multiplying and Dividing Fractions

Station 2

You will use strips of paper to help you divide fractions. Fold a strip of paper in half four times, without opening it between folds. Open the strip of paper. You should have 16 equal sections.

To solve the problem $\dfrac{3}{8} \div \dfrac{1}{16}$, first use a yellow highlighter to color $\dfrac{3}{8}$ of the strip. Work together to make sure the correct number of sections are colored.

Then use a blue highlighter to color $\dfrac{1}{16}$ of the strip. (All the sections that you color blue should be within the yellow section.) Work together to make sure the correct number of sections are colored.

Find the number of times the blue section would fit inside the yellow section. This is the quotient. Write it below.

_____

Use the above method to solve these problems.

1. $\dfrac{1}{2} \div \dfrac{1}{8} =$ _____

2. $\dfrac{3}{4} \div \dfrac{1}{16} =$ _____

3. $1 \div \dfrac{1}{4} =$ _____

4. $\dfrac{3}{4} \div \dfrac{3}{8} =$ _____

5. $\dfrac{5}{8} \div \dfrac{5}{16} =$ _____

The Number System
Set 4: Multiplying and Dividing Fractions

Station 3

You will be given two sets of cards labeled "A" and "B." Choose one A card and one B card.

Turn the cards over. Divide the fraction on card A by the fraction on card B. Write the division problem below.

_____

Work as a group to divide the fractions. Everyone in the group should agree on your answer. Write the answer below.

_____

Put the cards back. Mix up the cards. Repeat the above process three more times.

Division problem: _____

Answer: _____

Division problem: _____

Answer: _____

Division problem: _____

Answer: _____

Explain the strategies you used to solve the division problems.

_____

_____

_____

_____

_____

The Number System
Set 4: Multiplying and Dividing Fractions

Station 4

You will find a set of cards and a penny at this station. You will use these to create multiplication and division problems involving fractions and mixed numbers.

Choose two cards without looking. Turn the cards over. Write the fractions or mixed numbers in the two boxes below. Then flip the penny. If the penny lands heads up, write "×" on the line between the boxes. If it lands tails up, write "÷" on the line.

Work with other students to find the product or quotient of the fractions or mixed numbers. Simplify the answer if possible. When everyone agrees on the answer, write it below.

_____

Repeat the process four more times.

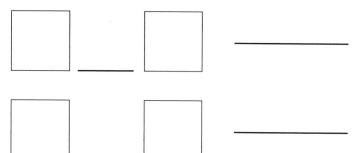

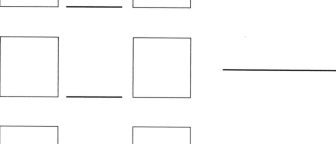

Expressions and Equations

Set 1: Graphing Relationships

Goal: To provide opportunities for students to develop concepts and skills related to coordinate graphing

Common Core Standards

Expressions and Equations

Apply and extend previous understandings of arithmetic to algebraic expressions.

6.EE.2. Write, read, and evaluate expressions in which letters stand for numbers.

Represent and analyze quantitative relationships between dependent and independent variables.

6.EE.9. Use variables to represent two quantities in a real-world problem that change in relationship to one another; write an equation to express one quantity, thought of as the dependent variable, in terms of the other quantity, thought of as the independent variable. Analyze the relationship between the dependent and independent variables using graphs and tables, and relate these to the equation.

Student Activities Overview and Answer Key

Station 1

Students work together to complete a table for the relationship $y = 3x$. They use the completed table to write ordered pairs that satisfy the equation and then work together to plot these points on a coordinate plane. Once students have plotted the points, they complete the graph and describe it.

Answers

1. See the following page for table. 2. (0, 0), (1, 3), (2, 6), (3, 9), (4, 12); 3–5. See the following page for graph. 6. Possible answer: The graph is a straight line that passes through the origin.

x	y
0	0
1	3
2	6
3	9
4	12

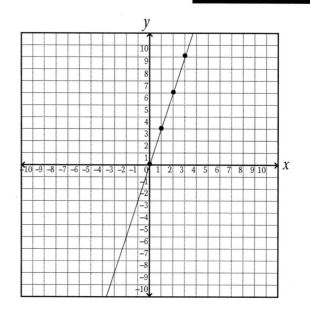

Station 2

In this activity, students are given a table of values. They work together to plot these points and look for patterns. (The points all satisfy the relationship $y = 1.5x$, so they all lie on a straight line.) Using the graph of plotted points, students name an additional point that they think satisfies the same relationship and they justify their choice.

Answers

1. (–4, –6), (–2, –3), (1, 1.5), (4, 6), (5, 7.5); 2–3. See below for graph. 4. The points all lie on a straight line. 5. Possible answer: (–6, –9) or (2, 3) 6. The point lies along the line determined by the other points.

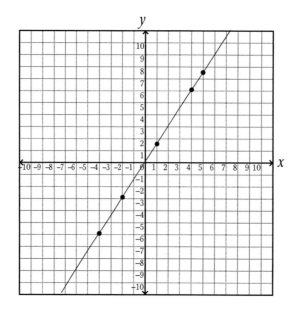

Expressions and Equations
Set 1: Graphing Relationships

Station 3

Students work together to explore a pattern made from tiles. They complete a table showing the relationship between the stage of the pattern and the number of tiles needed. Then they plot the points (which lie on the line $y = 4x$) and use the graph to predict the number of tiles that would be needed to make Stage 6 of the pattern.

Answers

1. See below for table. 2. (1, 4), (2, 8), (3, 12), (4, 16), (5, 20); 3–4. See below for graph.
5. 24 tiles; 6. Possible explanation: The plotted points all lie on a line. Continue the pattern of points to find that the next one is (6, 24).

Stage (x)	1	2	3	4	5
Number of tiles (y)	4	8	12	16	20

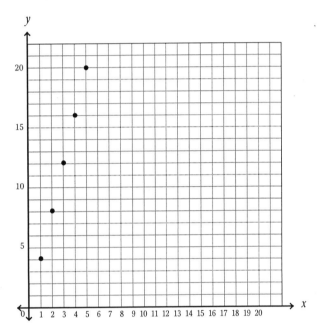

Station 4

Students work together to complete a table for the relationship $y = -2x$. They use the completed table to write ordered pairs that satisfy the equation and then work together to plot these points on a coordinate plane. Once students have plotted the points, they complete the graph and describe it.

Answers

1. See the following page for table. 2. (–3, 6), (–1, 2), (0, 0), (2, –4), (4, –8); 3–5. See the following page for graph. 6. Possible answer: The graph is a downward-sloping straight line that passes through the origin.

x	y
−3	6
−1	2
0	0
2	−4
4	−8

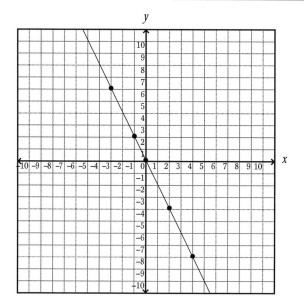

Materials List/Setup

Station 1   graph paper; ruler

Station 2   graph paper; ruler

Station 3   small square tiles; graph paper; ruler

Station 4   graph paper; ruler

Discussion Guide

To support students in reflecting on the activities, and to gather formative information about student learning, use the following prompts to facilitate a class discussion to "debrief" the station activities.

Prompts/Questions

1. How do you set up a coordinate plane on a sheet of graph paper?

2. How do you plot an ordered pair on a coordinate plane?

3. How do you get ordered pairs from a table of values?

4. What can you say about the graph of any equation that has the form $y = kx$?

Think, Pair, Share

Have students jot down their own responses to questions, discuss their responses with a partner (who was not in their station group), and then discuss as a whole class.

Suggested Appropriate Responses

1. Use a ruler to draw a straight line (*x*-axis). Draw another line at a right angle to the first line (*y*-axis). Label the intersection as the origin, (0, 0).

2. First plot the *x*-value by moving along the *x*-axis by the given number of units. Then move up (or down) from this point to plot the *y*-value.

3. In each row or column, take the *x*-value and the corresponding *y*-value, in that order, to form one ordered pair.

4. The graph is a straight line that passes through the origin.

Possible Misunderstandings/Mistakes

- Plotting points incorrectly due to reversing the *x*- and *y*-values (e.g., plotting (3, 4) rather than (4, 3))

- Plotting points incorrectly due to mislabeling or incorrectly calibrating the scale on the *x*- or *y*-axis

- Incorrectly reading the ordered pairs from a table of values

Expressions and Equations

Set 1: Graphing Relationships

Station 1

At this station, you will work together to make a table and graph for the relationship $y = 3x$.

The equation $y = 3x$ states that the value of y is 3 times the value of x.

x	y
0	
1	
2	6
3	
4	

1. Work with other students to complete the table of values.

2. Write the ordered pairs of values (x, y) from your table.

_____

3. Set up a coordinate plane on a sheet of graph paper. Use a ruler to draw the x-axis and the y-axis.

4. Plot the ordered pairs from your table. Work together to make sure that all the points are plotted correctly.

5. Use the points you plotted to draw the complete graph of $y = 3x$.

6. Describe the graph of $y = 3x$.

_____

_____

_____

_____

NAME: _____

Expressions and Equations

Set 1: Graphing Relationships

Station 2

At this station, you will plot points from a table and use your graph to make predictions.

The table to the right shows a set of values. Work with other students to explore how the *x*- and *y*-values are related.

x	*y*
−4	−6
−2	−3
1	1.5
4	6
5	7.5

1. Write the ordered pairs of values, (*x*, *y*), from the table.

   _____

2. Set up a coordinate plane on a sheet of graph paper. Use a ruler to draw the *x*-axis and the *y*-axis.

3. Plot the ordered pairs. Work together to make sure that all the points are plotted correctly.

4. Describe what you notice about the points.

   _____

   _____

   _____

5. Add a new point to the graph that you think has the same relationship between the *x*-value and *y*-value. What are the coordinates of the point? _____

6. Explain how you decided on this point.

   _____

   _____

   _____

Expressions and Equations
Set 1: Graphing Relationships

Station 3

At this station, you will use a table and graph to explore a pattern. You can also use tiles to build the pattern.

Marissa is using tiles to make a pattern. Here are the first three stages in her pattern.

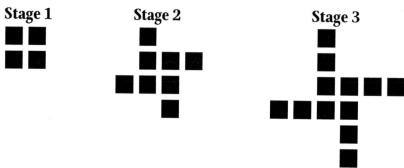

1. Complete the table.

Stage (x)	1	2	3	4	5
Number of tiles (y)					

2. Write the ordered pairs of values (x, y) from the table.

   _____

3. Set up a coordinate plane on a sheet of graph paper. Use a ruler to draw the x-axis and the y-axis.

4. Plot the ordered pairs. Work together to make sure that all the points are plotted correctly.

5. Use your graph to predict the number of tiles that are needed at Stage 6. _____

6. Explain how you made this prediction.

   _____

   _____

   _____

Expressions and Equations
Set 1: Graphing Relationships

Station 4

At this station, you will work together to make a table and graph for the relationship $y = -2x$.

The equation $y = -2x$ states that the value of y is -2 times the value of x.

x	y
-3	
-1	
0	
2	-4
4	

1. Work with other students to complete the table of values.

2. Write the ordered pairs of values from your table.

_____

3. Set up a coordinate plane on a sheet of graph paper. Use a ruler to draw the x-axis and the y-axis.

4. Plot the ordered pairs from your table. Work together to make sure that all the points are plotted correctly.

5. Use the points you plotted to draw the complete graph of $y = -2x$.

6. Describe the graph of $y = -2x$.

_____

_____

_____

_____

Expressions and Equations

Goal: To provide opportunities for students to develop concepts and skills related to evaluating expressions

Common Core Standards

Expressions and Equations

Apply and extend previous understandings of arithmetic to algebraic expressions.

6.EE.2. Write, read, and evaluate expressions in which letters stand for numbers.

    a. Write expressions that record operations with numbers and with letters standing for numbers.

    c. Evaluate expressions at specific values of their variables. Include expressions that arise from formulas used in real-world problems. Perform arithmetic operations, including those involving whole-number exponents, in the conventional order when there are no parentheses to specify a particular order (Order of Operations).

Reason about and solve one-variable equations and inequalities.

6.EE.5. Understand solving an equation or inequality as a process of answering a question: which values from a specified set, if any, make the equation or inequality true? Use substitution to determine whether a given number in a specified set makes an equation or inequality true.

Student Activities Overview and Answer Key

Station 1

In this activity, students work together to evaluate expressions. They choose an expression at random from a set of cards and then roll a number cube to get a value for the variable. Then they work together to evaluate the expression for this value of the variable. After repeating the process several times, students reflect on the strategies they used to evaluate the expressions.

Answers

Answers will depend upon the cards that are chosen and the numbers that are rolled.

Possible strategies: Substitute the value of the variable in the expression. Simplify the numerical expression using the order of operations.

Station 2

Students work together to evaluate the expression $2x + 1$ for a variety of values of x. They also choose their own values of x to substitute into the expression. After doing so, students look back on their results and look for a pattern. They find that no matter what counting number is used as the value of x, the result of evaluating $2x + 1$ is an odd number.

Answers

For $x = 1$, $2x + 1 = 3$; for $x = 2$, $2x + 1 = 5$; for $x = 3$, $2x + 1 = 7$; for $x = 7$, $2x + 1 = 15$; for $x = 9$, $2x + 1 = 19$. Other values of the expression will depend upon values chosen for the variable. No matter what counting number is used for the value of x, the value of $2x + 1$ is always odd.

Station 3

Students work together to complete tables by evaluating expressions for given values of the variable. Then students describe any patterns they see in the tables. In each case, the values of the variable increase by 1, while the values of the expression increase by a constant amount.

Answers

Value of x	1	2	3	4	5
Value of $3x + 5$	8	11	14	17	20

Possible patterns: The values of x increase by 1; the values of $3x + 5$ increase by 3.

Value of x	3	4	5	6	7
Value of $10x - 2$	28	38	48	58	68

Possible patterns: The values of x increase by 1; the values of $10x - 2$ increase by 10.

Station 4

Students work together to match a set of given expressions with a set of values of the variable. The goal is to pair expressions and values of the variable so that every expression has a value of 12 when evaluated for the value of the variable with which it is paired. All students should agree on the pairing of the cards before writing the answer.

Answers

The cards should be paired as follows: $2x + 2$ and $x = 5$; $5 + x$ and $x = 7$; $36 \div x$ and $x = 3$; $16 - x$ and $x = 4$; $3x^2$ and $x = 2$.

Possible strategies: Choose an expression and evaluate it for each possible value of the variable until the result is 12. Alternatively, choose an expression and decide which value of the variable makes it equal 12, and then match it with this value.

Expressions and Equations
Set 2: Evaluating Expressions

Materials List/Setup

Station 1   number cube (numbers 1–6)
set of 5 index cards with the following expressions written on them:
$3n + 4, 2m^2, 15 - x, 60 \div p, 2t - 2$

Station 2   none

Station 3   none

Station 4   set of 5 index cards with the following expressions written on them:
$2x + 2, 5 + x, 36 \div x, 16 - x, 2m^2$

set of 5 index cards with the following values of x written on them:
$x = 2, x = 3, x = 4, x = 5, x = 7$

Discussion Guide

To support students in reflecting on the activities, and to gather formative information about student learning, use the following prompts to facilitate a class discussion to "debrief" the station activities.

Prompts/Questions

1. How do you evaluate an expression for a given value of the variable?

2. After you substitute the value of the variable in the expression, how do you simplify the result?

3. How do you evaluate the expression $2n^2$ for a specific value of n?

4. What strategies can you use to help you find patterns in a table of values?

Think, Pair, Share

Have students jot down their own responses to questions, discuss their responses with a partner (who was not in their station group), and then discuss as a whole class.

Suggested Appropriate Responses

1. Substitute the value of the variable for the variable in the expression and simplify.

2. Perform the operation(s), using the order of operations if necessary.

3. Substitute the value for n. Then square the value by multiplying it by itself. Then multiply by 2.

4. Look for patterns as you go from one value to the next. For example, check to see if the values increase or decrease by a fixed number each time.

Possible Misunderstandings/Mistakes

- Applying an incorrect operation (e.g., adding instead of multiplying when evaluating for a value of x)

- Evaluating exponents incorrectly

- Incorrectly applying the order of operations when evaluating an expression that involves more than one operation

Expressions and Equations
Set 2: Evaluating Expressions

Station 1

At this station, you will find a stack of cards and a number cube. Each card contains an algebraic expression.

Choose one of the cards without looking. Write the expression on the line. _____

Roll the number cube to get a value for the variable. Write this value on the line.

_____

Work with other students to evaluate the expression for this value of the variable. When everyone agrees on the result, write it below.

_____

Put the card back. Mix up the cards. Repeat the above process four more times.

      Expression on card: _____

      Value of variable: _____

      Evaluate the expression: _____

      Expression on card: _____

      Value of variable: _____

      Evaluate the expression: _____

      Expression on card: _____

      Value of variable: _____

      Evaluate the expression: _____

      Expression on card: _____

      Value of variable: _____

      Evaluate the expression: _____

Describe the strategies you used to evaluate the expressions for the given values of the variable.

_____

_____

_____

Expressions and Equations
Set 2: Evaluating Expressions

Station 2

At this station, you will work with other students to evaluate expressions and look for patterns.

Work together to evaluate the expression $2x + 1$ for each value of the variable shown below. When everyone agrees on an answer, write it down.

$x = 1$ _____

$x = 2$ _____

$x = 3$ _____

$x = 7$ _____

$x = 9$ _____

Choose your own three values for x. (The values should be counting numbers, such as 1, 2, 3, and so on.) Then evaluate the expression $2x + 1$ for these values of x.

Value of x : _____

Evaluate the expression: _____

Value of x : _____

Evaluate the expression: _____

Value of x : _____

Evaluate the expression: _____

No matter what counting number you substitute for x, what is true about the result when you evaluate $2x + 1$?

_____

_____

_____

Expressions and Equations
Set 2: Evaluating Expressions

Station 3

At this station, you will work with other students to complete tables and look for patterns.

Work with other students to complete this table. To do so, evaluate $3x + 5$ for each value of x in the table.

Value of x	1	2	3	4	5
Value of $3x + 5$					

Look for patterns in the table. Describe at least two patterns that you see.

_____

_____

_____

_____

Work together to complete this table. This time, evaluate $10x - 2$ for each value of x in the table.

Value of x	3	4	5	6	7
Value of $10x - 2$					

Look for patterns in the table. Describe at least two patterns that you see.

_____

_____

_____

_____

Expressions and Equations

Set 2: Evaluating Expressions

Station 4

At this station, you will find five cards with these expressions written on them.

$$2x + 2 \qquad 5 + x \qquad 36 \div x \qquad 16 - x \qquad 3x^2$$

You will also find five cards with these values of x written on them.

$$x = 2 \qquad x = 3 \qquad x = 4 \qquad x = 5 \qquad x = 7$$

Work together to match each expression with a value of x. When you evaluate each expression for its value of x, the result should be 12.

Work together to check that each pair gives a value of 12.

When everyone agrees on the results, write the five pairs below.

_____

_____

_____

_____

_____

Describe the strategies you used to solve this problem.

_____

_____

_____

_____

Expressions and Equations

Set 3: Solving Equations

Goal: To provide opportunities for students to develop concepts and skills related to
solving equations

Common Core Standards

Expressions and Equations

Apply and extend previous understandings of arithmetic to algebraic expressions.

6.EE.2. Write, read, and evaluate expressions in which letters stand for numbers.

Reason about and solve one-variable equations and inequalities.

6.EE.5. Understand solving an equation or inequality as a process of answering a question: which values from a specified set, if any, make the equation or inequality true? Use substitution to determine whether a given number in a specified set makes an equation or inequality true.

6.EE.6. Use variables to represent numbers and write expressions when solving a real-world or mathematical problem; understand that a variable can represent an unknown number, or, depending on the purpose at hand, any number in a specified set.

Student Activities Overview and Answer Key

Station 1

Students are given a set of cards with simple one-step equations written on them. They are given another set of cards with values of the variable written on them. Students work together to match each equation to its solution. Once students have paired the cards, they reflect on the strategies they used.

Answers

The cards should be paired as follows: $8 + x = 12$ and $x = 4$; $x/4 = 2$ and $x = 8$; $6x = 18$ and $x = 3$; $x - 4 = 2$ and $x = 6$; $14 = 7x$ and $x = 2$; $2 = x/6$ and $x = 12$.

Possible strategies: Choose an equation. Check each value of x in the equation to see if it is a solution. Alternatively, solve the equation and look for its solution among the values of x.

Station 2

Students work together to use algebra tiles to represent simple one-step equations. Then they use the tiles to help them solve the equations. Students explain the strategies they used to manipulate the tiles and solve the equations.

Answers

1. $x = 3$; 2. $x = 6$; 3. $x = 2$; 4. $x = 2$; 5. $x = 6$; 6. $x = 3$

Possible strategies: Remove the same number of yellow tiles from both sides of the mat. Divide the tiles on each side of the mat into the same number of equal groups, and remove all but one of the groups on each side. The remaining group shows the solution.

Station 3

In this activity, students use cups and counters to model simple one-step equations. In the given pictures, each cup is holding an unknown number of counters. Students use this idea to write the equation that is modeled by each picture. Then they use actual cups and counters, as well as logical reasoning, to help them find the unknown number of counters in each cup. This is equivalent to solving the corresponding equation.

Answers

1. $x + 2 = 9$, $x = 7$; 2. $6 = x + 1$, $x = 5$; 3. $2x = 10$, $x = 5$; 4. $3x = 12$, $x = 4$

Station 4

Students are given a set of equations and a set of real-world situations. They work together to match each situation to an equation. Then they solve the equation. At the end of the activity, students explain the strategies they used to match the equations to the situations.

Answers

1. $x + 6 = 18$, $x = 12$; 2. $x/6 = 18$, $x = 108$; 3. $6x = 18$, $x = 3$; 4. $x - 6 = 18$, $x = 24$

Possible strategies: Use the words or phrases that refer to arithmetic operations as clues to identifying the corresponding equations. For example, separating into equal groups corresponds to division.

Materials List/Setup

Station 1	set of 6 index cards with the following equations written on them: $8 + x = 12$, $x/4 = 2$, $6x = 18$, $x - 4 = 2$, $14 = 7x$, $2 = x/6$ set of 6 index cards with the following values of x written on them: $x = 2$, $x = 3$, $x = 4$, $x = 6$, $x = 8$, $x = 12$
Station 2	algebra tiles and equation mat
Station 3	3 paper cups; 12 counters or other small objects, such as pennies or beans
Station 4	none

Discussion Guide

To support students in reflecting on the activities, and to gather formative information about student learning, use the following prompts to facilitate a class discussion to "debrief" the station activities.

Prompts/Questions

1. What are some different tools, objects, or drawings that you can use to help you solve equations?

2. What does the equal sign tell you in an equation?

3. How do you solve an equation using inverse operations?

4. How can you check your solution to an equation?

Think, Pair, Share

Have students jot down their own responses to questions, discuss their responses with a partner (who was not in their station group), and then discuss as a whole class.

Suggested Appropriate Responses

1. algebra tiles, cups and counters, drawings of balance scales, etc.

2. It tells you that the quantities on either side of the equation are the same.

3. Isolate the variable by applying inverse operations to both sides of the equation.

4. Substitute the value for the variable in the equation and simplify. If the solution is correct, the two sides of the equation should be equal.

Possible Misunderstandings/Mistakes

- Using an incorrect operation to solve an equation (e.g., solving $x + 3 = 12$ by adding 3 to both sides)

- Attempting to solve an equation such as $x + 4 = 9$ by subtracting x from both sides

- Applying an operation that does not isolate the variable (e.g., solving $9 = 3x$ by dividing both sides by 9)

Expressions and Equations
Set 3: Solving Equations

Station 1

At this station, you will find a set of cards with the following equations written on them:

$$8 + x = 12 \qquad \frac{x}{4} = 2 \qquad 6x = 18 \qquad x - 4 = 2 \qquad 14 = 7x \qquad 2 = \frac{x}{6}$$

You will also find a set of cards with the following values of x written on them:

$$x = 2 \qquad x = 3 \qquad x = 4 \qquad x = 6 \qquad x = 8 \qquad x = 12$$

Work with other students to match each equation with its solution.

Work together to check that each equation is paired with its correct solution. Write the pairs below.

_____

_____

_____

_____

_____

_____

Explain the strategies you used to match up the cards.

_____

_____

_____

_____

Expressions and Equations
Set 3: Solving Equations

Station 2

You can use algebra tiles to help you solve equations.

Each square yellow tile shows +1. Each square red tile shows –1. Each rectangular yellow tile shows x. You use the equation mat to show the two sides of an equation.

Work together to use algebra tiles to show each equation. Then use the tiles to solve each equation. Write the value of x for each equation below.

1. $x + 5 = 8$ _____

2. $9 = x + 3$ _____

3. $4 + x = 6$ _____

4. $3x = 6$ _____

5. $2x = 12$ _____

6. $9 = 3x$ _____

Explain at least two strategies you used to solve the equations using algebra tiles.

_____

_____

_____

_____

Expressions and Equations
Set 3: Solving Equations

Station 3

In each picture, the cup is holding an unknown number of counters, x. If there is more than one cup, every cup is holding the same number of counters.

Each picture shows an equation. This picture shows $x + 5 = 7$. To make the two sides equal, there must be 2 counters in the cup. This means $x = 2$.

Work with other students to write an equation for each picture. Then find the number of counters in each cup. You can use the cups and counters at the station to help you.

1.
   Equation: _____

   Solution: _____

2.
   Equation: _____

   Solution: _____

3.
   Equation: _____

   Solution: _____

4.
   Equation: _____

   Solution: _____

Expressions and Equations
Set 3: Solving Equations

Station 4

At this station, you will match equations to real-world situations and then solve the equations.

Work with other students to match each situation to one of the following equations. When everyone agrees on the correct equation, write it in the space provided. Then work together to solve it.

$$\frac{x}{6} = 18 \qquad x - 6 = 18 \qquad 6x = 18 \qquad x + 6 = 18$$

1. In 6 years, Rosario will be 18 years old. How old is she now?

   Equation: _____

   Solution: _____

2. When the students at Essex Middle School are separated into 6 equal groups, there are 18 students in each group. How many students are at the school?

   Equation: _____

   Solution: _____

3. Mike bought some books that cost $6 each. He spent a total of $18. How many books did he buy?

   Equation: _____

   Solution: _____

4. From 4:00 to 5:00 P.M., the temperature dropped 6°F. At 5:00 P.M. the temperature was 18°F. What was the temperature at 4:00 P.M.?

   Equation: _____

   Solution: _____

Explain the strategies you used to match the equations to the situations.

_____

_____

_____

Expressions and Equations

Goal: To provide opportunities for students to develop concepts and skills related to evaluating and simplifying expressions

Common Core Standards

Expressions and Equations

Apply and extend previous understandings of arithmetic to algebraic expressions.

6.EE.3. Apply the properties of operations to generate equivalent expressions.

6.EE.4. Identify when two expressions are equivalent (i.e., when the two expressions name the same number regardless of which value is substituted into them).

Student Activities Overview and Answer Key

Station 1

In this activity, students work together to evaluate expressions from a real-life situation. They are given expressions for converting temperatures in degrees Celsius to degrees Fahrenheit and vice versa. Students work together to evaluate these expressions for given values, and then they reflect on the steps that they used.

Answers

1. 50; 2. 77; 3. 212; 4. 23; 5. −4

Possible steps: Substitute the given value for C. Divide this value by 5 and multiply by 9. Then add 32.

6. 20; 7. 0; 8. −5; 9. −20; 10. −40

Possible steps: Substitute the given value for F. Subtract 32 from this value. Then divide the result by 9 and multiply by 5.

Station 2

Students are given a set of cards with algebraic expressions written on them. Students work together to sort the cards into pairs so that the cards in each pair show equivalent expressions. Then they explain the strategies they used to solve the problem.

Answers

The cards should be paired as follows: $3x^2 + 4x + 4x^2$ and $7x^2 + 4x$, $5x + 2x + x^2$ and $7x + x^2$, $2(x + x^2) + 3x$ and $5x + 2x^2$, $8x + 6x^2 − 3x − x^2$ and $5x + 5x^2$, $x^2 + 3(x^2 + 2)$ and $4x^2 + 6$.

Possible strategies: First simplify the expressions on the cards that have expressions that can be simplified. Then look for another card that contains the simplified expression.

Station 3

Students are given a set of cards with simple algebraic expressions on them. Students choose three cards at random and write the expressions in boxes provided on the activity sheet. In this way, students generate expressions that they will simplify by working together.

Answers

Answers will depend upon the cards that are chosen.

Station 4

Students work together to match a set of given expressions with a set of integer values of the variable. The goal is to pair expressions and values of the variable so that every expression has a value of 12 when evaluated for the value of the variable with which it is paired. All students should agree on the pairing of the cards before writing the answer.

Answers

The cards should be paired as follows: $-2x + 2$ and $x = -5$, $16 + x$ and $x = -4$, $-36 \div x$ and $x = -3$, $20 - 2x$ and $x = 4$, and $3x^2$ and $x = -2$.

Possible strategies: Choose an expression and evaluate it for each possible value of the variable until the result is 12. Alternatively, choose an expression and decide which value of the variable makes it equal 12, then match it with this value.

Materials List/Setup

Station 1	none
Station 2	set of 10 index cards with the following expressions written on them: $3x^2 + 4x + 4x^2$, $7x + x^2$, $5x + 5x^2$, $8x + 6x^2 - 3x - x^2$, $5x + 2x + x^2$, $7x^2 + 4x$, $5x + 2x^2$, $x^2 + 3(x^2 + 2)$, $2(x + x^2) + 3x$, $4x^2 + 6$
Station 3	set of 10 index cards with the following expressions written on them: y, $2y$, $3y$, $6y$, $9y$, y^2, $3y^2$, $4y^2$, $6y^2$, $8y^2$
Station 4	set of 5 index cards with the following expressions written on them: $-2x + 2$, $16 + x$, $-36 \div x$, $20 - 2x$, $3x^2$ set of 5 index cards with the following values of x written on them: $x = -2$, $x = -3$, $x = -4$, $x = -5$, $x = 4$

Discussion Guide

To support students in reflecting on the activities and to gather some formative information about student learning, use the following prompts to facilitate a class discussion to "debrief" the station activities.

Prompts/Questions

1. How do you evaluate an expression for a given value of the variable?

2. After you substitute the value of the variable in the expression, how do you simplify the result?

3. How do you evaluate the expression $12n^2 + 1$ for a specific value of n?

4. What steps do you use to simplify an algebraic expression?

Think, Pair, Share

Have students jot down their own responses to questions, then discuss with a partner (who was not in their station group), and then discuss as a whole class.

Suggested Appropriate Responses

1. Substitute the value of the variable for the variable in the expression and simplify.

2. Perform the operation(s), using the order of operations if necessary.

3. Substitute the value for n. Then square the value by multiplying it by itself. Then multiply by 12. Then add 1.

4. Look for like terms. Combine the like terms by adding or subtracting coefficients.

Possible Misunderstandings/Mistakes

- Applying an incorrect operation (e.g., adding instead of multiplying when evaluating $6x$ for a value of x)

- Incorrectly applying the order of operations when evaluating or simplifying an expression that involves more than one operation

- Incorrectly applying the Distributive Property (e.g., writing $3(x + 1) = 3x + 1$)

Expressions and Equations
Set 4: Evaluating and Simplifying Expressions

Station 1

Given a temperature C in degrees Celsius, the expression $\frac{9}{5}C + 32$ gives the temperature in degrees Fahrenheit.

Work together to evaluate the expression $\frac{9}{5}C + 32$ for the following values of C.

1. $C = 10$ _____
2. $C = 25$ _____
3. $C = 100$ _____
4. $C = -5$ _____
5. $C = -20$ _____

Explain the steps you used to evaluate these expressions. _____

_____

_____

Given a temperature F in degrees Fahrenheit, the expression $\frac{5}{9}(F - 32)$ gives the temperature in degrees Celsius.

Work together to evaluate the expression $\frac{5}{9}(F - 32)$ for the following values of F.

6. $F = 68$ _____
7. $F = 32$ _____
8. $F = 23$ _____
9. $F = -4$ _____
10. $F = -40$ _____

Explain the steps you used to evaluate these expressions. _____

_____

_____

Expressions and Equations
Set 4: Evaluating and Simplifying Expressions

Station 2

You will find a set of 10 cards at this station. The cards contain the following expressions:

$$3x^2 + 4x + 4x^2 \qquad\qquad 7x^2 + 4x$$

$$7x + x^2 \qquad\qquad 5x + 2x^2$$

$$5x + 5x^2 \qquad\qquad x^2 + 3(x^2 + 2)$$

$$8x + 6x^2 - 3x - x^2 \qquad\qquad 2(x + x^2) + 3x$$

$$5x + 2x + x^2 \qquad\qquad 4x^2 + 6$$

Work together to sort the cards into pairs. The cards in each pair should show equivalent expressions.

When everyone agrees on the answer, write the five pairs below.

_____

_____

_____

_____

_____

Explain the strategies you used to solve this problem. _____

_____

_____

_____

Expressions and Equations
Set 4: Evaluating and Simplifying Expressions

Station 3

You will find a set of cards at this station. The cards should be spread out, face down.

Choose three cards without looking. Write the expressions on the cards in the boxes below.

Work together to simplify the expression. When everyone agrees on the answer, write it below.

Simplified expression: _____

Put the cards back. Mix up the cards. Then repeat the above process four more times.

Simplified expression: _____

Simplified expression: _____

Simplified expression: _____

Simplified expression: _____

Station Activities for Common Core Mathematics, Grade 6

Expressions and Equations

Set 4: Evaluating and Simplifying Expressions

Station 4

At this station, you will find five cards with the following expressions written on them:

$$-2x + 2 \qquad 16 + x \qquad -36 \div x \qquad 20 - 2x \qquad 3x^2$$

You will also find five cards with the following values of x written on them:

$$x = -2 \qquad x = -3 \qquad x = -4 \qquad x = -5 \qquad x = 4$$

Work together to match each expression with a value of x. When you evaluate each expression for its value of x, the result should be 12.

Work together to check that each pair gives a value of 12.

When everyone agrees on the results, write the five pairs below.

_____

_____

_____

_____

_____

Describe the strategies you used to solve this problem. _____

_____

_____

_____

Expressions and Equations

Goal: To provide opportunities for students to develop concepts and skills related to
solving inequalities

Common Core Standards

Expressions and Equations

Reason about and solve one-variable equations and inequalities.

6.EE.8. Write an inequality of the form $x > c$ or $x < c$ to represent a constraint or condition in a
real-world or mathematical problem. Recognize that inequalities of the form $x > c$ or
$x < c$ have infinitely many solutions; represent solutions of such inequalities on
number line diagrams.

Student Activities Overview and Answer Key

Station 1

Students are given a series of inequalities and a number cube. For each inequality, they roll the
number cube and then work together to decide if the number shown on the cube is a solution of the
inequality. Students explain the strategies they used to decide whether each value was a solution.

Answers

1–3. Answers will depend upon numbers rolled. 4. yes; 5. no

Possible strategies: Substitute the value for the variable. Simplify and check to see if the resulting
inequality is true.

Station 2

In this activity, students work together to use number lines to help them solve inequalities. To do so,
they test various values of the variable in the given inequalities, and check to see whether each value
is a solution. They keep track of the values that are solutions by marking them on a number line.
After testing enough values to see a pattern, students shade the values that represent all solutions of
the inequality. Then they write the solution algebraically.

Answers

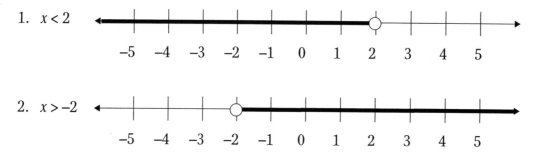

1. $x < 2$

2. $x > -2$

3. $x < 1$

4. $x < 1$

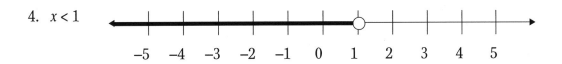

5. $x < 2$

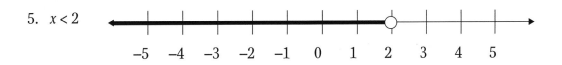

Station 3

In this activity, students work together to match a set of given inequalities with a set of given solutions. Once students have paired each inequality with its correct solution, they discuss the strategies they used to solve the problem.

Answers

The cards should be paired as follows: $5x + 2 < 12$ and $x < 2$; $4x + 3 < -5$ and $x < -2$; $-3x < 6$ and $x > -2$; $-x/_2 > 2$ and $x < -4$; $3x + 1 > -11$ and $x < -4$; $x/_4 + 1 > 2$ and $x > 4$.

Possible strategies: Solve each inequality using inverse operations and look for the solution among the given choices.

Station 4

Students are given a set of inequalities and a set of real-world situations. They work together to match each situation to an inequality. Then they solve the inequality. At the end of the activity, students explain the strategies they used to match the inequalities to the situations.

Answers

1. $10x + 5 < 105$, $x < 11$; 2. $5x + 10 < 105$, $x < 19$; 3. $10x - 5 > 105$, $x > 11$; 4. $5(x - 10) < 105$, $x < 31$

Possible strategies: Use the words or phrases that refer to arithmetic operations as clues to identifying the corresponding inequalities. For example, a decrease corresponds to subtraction. Match words to inequalities. For example, "no more than but not including" refers to a "less than" inequality (<).

Materials List/Setup

Station 1   number cube (numbers 1–6)

Station 2   none

Station 3   set of index cards with the following inequalities written on them:
$5x + 2 < 12$, $4x + 3 < -5$, $-3x < 6$, $-x/2 > 2$, $3x + 1 > -11$, $x/4 + 1 > 2$
set of index cards with the following solutions written on them:
$x < -4$, $x > -4$, $x < -2$, $x > -2$, $x < 2$, $x > 4$

The two sets of cards should be placed in two piles, face-up, on a table or desk at the station.

Station 4   none

Discussion Guide

To support students in reflecting on the activities and to gather some formative information about student learning, use the following prompts to facilitate a class discussion to "debrief" the station activities.

Prompts/Questions

1. What is the difference between < and >?

2. How do you check to see if a value of the variable is a solution of an inequality?

3. How is the solution of an inequality different from the solution of an equation?

4. How do you solve an equality using algebra?

Think, Pair, Share

Have students jot down their own responses to questions, then discuss with a partner (who was not in their station group), and then discuss as a whole class.

Suggested Appropriate Responses

1. The symbol < means "less than." The symbol > means "more than."

2. Substitute the value for the variable in the inequality. Check to see if the resulting inequality is true. If it is, the value is a solution.

3. In general, the solution of an inequality is itself an inequality (a range of values). The solution of an equation is usually a single value (or several discrete values).

4. Use inverse operations, as when solving an equation, to isolate the variable on one side of the inequality. If you multiply or divide by a negative number, reverse the direction of the inequality.

Possible Misunderstandings/Mistakes

- Using an incorrect operation to solve an inequality (e.g., solving $x + 2 < 5$ by adding 2 to both sides)

- Incorrectly translating verbal expressions to inequalities (e.g., representing the phrase "less than" by > rather than <)

- Forgetting to reverse the direction of the inequality when multiplying or dividing by a negative number

Expressions and Equations

Set 5: Solving Inequalities

Station 1

You will find a number cube at this station.

For each inequality, roll the number cube and write the number in the box. Then work together to decide if this value of the variable is a solution of the inequality. Write "yes" or "no" on the line provided.

1. $2x + 1 < 7$ Solution? _____

2. $3x - 4 > 5$ Solution? _____

3. $-3x < -12$ Solution? _____

4. $\dfrac{x}{2} + 1 < 5$ Solution? _____

5. $1 - x > 0$ Solution? _____

6. Explain the strategies you used to decide whether each value was a solution of the inequality.

_____

_____

_____

Expressions and Equations
Set 5: Solving Inequalities

Station 2

You can use number lines to help you solve inequalities.

For each inequality, work together to test different values of the variable to see if they are solutions of the inequality. If a value is a solution, draw a solid dot at that value on the number line. Test at least five different values for each inequality.

When you think you know what the solution set of an inequality looks like, shade the correct part of the number line to show all the solutions.

Finally, write the solution in the space provided.

1. $2x - 3 < 1$

   Solution: _____

2. $3x + 1 > -5$

   Solution: _____

3. $-2x < -2$

   Solution: _____

4. $4x - 2 < 2$

   Solution: _____

5. $\dfrac{x}{2} + 3 < 4$

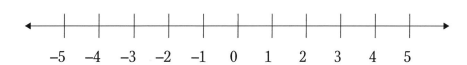

   Solution: _____

Expressions and Equations
Set 5: Solving Inequalities

Station 3

At this station, you will work with other students to match inequalities to their solutions.

You will find a set of cards with the following inequalities written on them:

$$5x + 2 < 12 \qquad 4x + 3 < -5 \qquad -3x < 6 \qquad -\frac{x}{2} > 2 \qquad 3x + 1 > -11 \qquad \frac{x}{4} + 1 > 2$$

You will also find a set of cards with these solutions written on them:

$$x < -4 \qquad x > -4 \qquad x < -2 \qquad x > -2 \qquad x < 2 \qquad x > 4$$

Work together to match each inequality with its solution. When everyone agrees on the answers, write the matching pairs below.

_____

_____

_____

_____

_____

_____

Explain the strategies you used to match up the cards.

_____

_____

_____

Expressions and Equations
Set 5: Solving Inequalities

Station 4

At this station, you will match inequalities to real-world situations and then solve the inequalities.

Work with other students to match each situation to one of the following inequalities. When everyone agrees on the correct inequality, write it on the line provided. Then work together to solve it.

$$5x + 10 < 105 \qquad 10x + 5 < 105 \qquad 5(x - 10) < 105 \qquad 10x - 5 > 105$$

1. Mai rents DVDs by mail. There is a one-time membership fee of $5 and the service costs $10 per month. Mai wants to spend less than $105. For how many months can she rent DVDs with this service?

    Inequality: _____

    Solution: _____

2. Tyrone bought 5 trays of food for a party. The price of each tray of food was the same. He also spent $10 on paper plates, napkins, and utensils. Altogether, he spent less than $105. What was the price of each tray of food?

    Inequality: _____

    Solution: _____

3. Mr. Garcia ordered 10 copies of a novel for students in his English class. He had a coupon for $5 off the total price of the order. The total cost of the order, before tax, came to more than $105. What was the price of each novel?

    Inequality: _____

    Solution: _____

continued

Expressions and Equations
Set 5: Solving Inequalities

4. Rachel bought 5 pairs of jeans. Each pair of jeans had the same price. She had a coupon for $10 off the price of each pair of jeans. The total cost of the jeans, before tax, came to less than $105. What was the price of each pair of jeans?

Inequality: _____

Solution: _____

Explain the strategies you used to match the inequalities to the situations.

_____

_____

_____

Geometry

Set 1: Appropriate Units of Measurement

Goal: To provide opportunities for students to develop concepts and skills related to using appropriate units for measurement

Common Core Standards

Geometry

Solve real-world and mathematical problems involving area, surface area, and volume.

6.G.2. ... show that the volume is the same as would be found by multiplying the edge lengths of the prism. Apply the formulas $V = lwh$ and $V = bh$ to find volumes of right rectangular prisms with fractional edge lengths in the context of solving real-world and mathematical problems.

Student Activities Overview and Answer Key

Station 1

At this station, students will measure a variety of rectangular prisms and find the volume of these objects. Students will use appropriate measuring techniques and units.

Answers

Answers will vary; inches; square inches; cubic inches; it depends on how many times you multiply inches

Station 2

Students will measure the length and width of the classroom to determine its area. They will explain their use of appropriate units.

Answers

Answers will vary; answers will vary; feet/yards because inches are too small; answers will vary; use the yard stick, choose units that were large, etc.

Station 3

Students find the approximate volume of their hand. They do this by finding the volume of six parts of the hand and adding that together. They then comment on their findings.

Answers

Answers will vary; cubic inches/centimeters, etc.; fingers/palms are not exactly rectangular prisms; answers may vary

Station 4

Students choose five objects in the classroom. They find the perimeter of these objects then reflect on the units that they chose to measure the objects in.

Answers

Answers will vary.

Materials List/Setup

Station 1	variety of items that are rectangular prisms; rulers
Station 2	yard stick; ruler
Station 3	rulers; calculators—enough for all group members
Station 4	yard stick; ruler

Discussion Guide

To support students in reflecting on the activities, and to gather formative information about student learning, use the following prompts to facilitate a class discussion to "debrief" the station activities.

Prompts/Questions

1. What is the difference between the units used in perimeter, area, and volume?

2. If you are measuring the side of an object using inches, and find that the length of the object is between 13 inches and 13 $\frac{1}{16}$, but closer to 13 $\frac{1}{16}$ inch, what do you use as the measure?

3. If you have two objects and one has a greater perimeter, does it have to have a greater area as well?

4. How is finding perimeter different from finding area?

Think, Pair, Share

Have students jot down their own responses to questions, discuss their responses with a partner (who was not in their station group), and then discuss as a whole class.

Suggested Appropriate Responses

1. perimeter = units, area = square units, volume = cubic units

2. The answer is 13 $\frac{1}{16}$ inches because we round to the nearest $\frac{1}{16}$ if that is the most precise our ruler can get.

3. no [e.g., 1×14 and 5×5. When comparing a rectangle measuring 1 foot $\times$ 14 foot with a rectangle measuring 5 feet $\times$ 5 feet, the 1×14 rectangle would have a perimeter of 30 feet and an area of 14 square feet, whereas the 5×5 rectangle would have a smaller perimeter (20 feet) but a larger area (25 square feet).]

4. We add all the lengths of the sides for perimeter but multiply when finding area.

Possible Misunderstandings/Mistakes

- Getting confused between $\frac{1}{4}$, $\frac{1}{8}$, and $\frac{1}{16}$ of an inch
- Difficulty measuring the hand
- Trouble deciding when it is better to use inches, feet, or yards

Geometry
Set 1: Appropriate Units of Measurement

Station 1

At this station, you will find a variety of objects and rulers. You will be finding the volume of these objects.

Each student should choose a different object. You will measure the length, width, and height of your object. Record the group data in the table below.

Group member	Object chosen	Length (in.)	Width (in.)	Height (in.)	Total volume

Which object has the greatest volume? _____

What units did you write the volume in? _____

How did you know what units to use? _____

_____

_____

Geometry
Set 1: Appropriate Units of Measurement

Station 2

Your goal at this station is to find the area of the room. You have a ruler and a yard stick to help you with this.

First measure the length and the width of the room and write them below.

Length (include units): _____

Width (include units): _____

Why did you choose the units you did? _____

_____

_____

What is the area of the room? (include units) _____

What was your strategy for measuring the room and deciding which units to use?

_____

_____

_____

_____

Geometry

Set 1: Appropriate Units of Measurement

Station 3

At this station, you will be finding the approximate volume of your hand. You will be using rulers and a calculator.

For this activity, think of your hand as six different rectangular prisms—the five fingers and your palm. Measure the length, width, and height of each of the six parts of the hand. It may be helpful to work with a group member for the measuring section of this activity.

Part of hand	Length	Width	Height	Volume
Pinky finger				
Ring finger				
Middle finger				
Index finger				
Thumb				
Palm				

What is the approximate volume of your hand? _____

What units did you use? Why? _____

_____

_____

Why might your results not be exact? _____

_____

_____

Which finger has the most volume? Does this surprise you? _____

_____

_____

Geometry
Set 1: Appropriate Units of Measurement

Station 4

At this station, you will find the perimeter of five different objects in the classroom. You will have a ruler and a yard stick to assist you.

First, choose five different objects to find the perimeter of (e.g., desk, room, etc).

Object	Length	Width	Perimeter

What object had the largest perimeter? _____

What units did you use? Were they always the same? _____

_____

_____

How did you decide what units to use? _____

_____

_____

© 2011 Walch Education

Geometry

Set 2: Visualizing Solid Figures

Goal: To provide opportunities for students to develop concepts and skills related to visualizing solid figures

Common Core Standards

Geometry

Solve real-world and mathematical problems involving area, surface area, and volume.

6.G.4. Represent three-dimensional figures using nets made up of rectangles and triangles, and use the nets to find the surface area of these figures. Apply these techniques in the context of solving real-world and mathematical problems.

Student Activities Overview and Answer Key

Station 1

Students will be drawing figures made from connecting cubes from different perspectives. They will then comment on their strategies for doing this.

Answers

Answers will vary; no—some are hidden by other blocks in some perspectives

Station 2

Students look at different possible nets for a cube. They decide whether or not they are, in fact, nets and then draw a net of their own.

Answers

yes; no; yes; yes; no; answers will vary

Station 3

Students will use connecting cubes to construct a figure based on a drawing. They then create and draw their own figure. Finally, students reflect on their strategies for completing the task.

Answers

15; yes—they are hidden behind other cubes; answers will vary

Station 4

Students draw nets for two different cylinders. They then compare and contrast these nets, and determine which cylinder has a greater surface area.

Answers

The rectangle is the same for both; the circles on the ends are different for both; the cylinder with the end circle whose circumference is 11

Materials List/Setup

Station 1   three different figures made up of 10 or fewer connecting cubes

Station 2   none

Station 3   25 connecting cubes

Station 4   tape; two 8 $\frac{1}{2}$" × 11" sheets of paper

Discussion Guide

To support students in reflecting on the activities and to gather some formative information about student learning, use the following prompts to facilitate a class discussion to "debrief" the station activities.

Prompts/Questions

1. When in real life do we have 2-D drawings of 3-D objects?

2. How many nets can you think of for a cube?

3. When would having a drawing on isometric dot paper be more useful than having a net for an object?

4. How can nets help us find surface area?

Think, Pair, Share

Have students jot down their own responses to questions, discuss their responses with a partner (who was not in their station group), and then discuss as a whole class.

Suggested Appropriate Responses

1. Many examples—topological maps, architectural elevations, etc.

2. 11

3. when you want to see depth

4. If we find the area of the net, we know the surface area of the object.

Possible Misunderstandings/Mistakes

- Not being able to visualize the nets—thinking they do or do not work when the reverse is true
- Not correctly drawing the figure they build
- Not being able to build the figure from the isometric dot paper—being confused by the perspective
- Confusing the front, side, and top drawings

Geometry
Set 2: Visualizing Solid Figures

Station 1

At this station, you will find connecting cubes that have been made into three different figures. You will use these figures to draw from different perspectives.

Split your group up into three subgroups and give each group one figure.

Draw the front perspective, the top perspective, and the right side perspective of the figures on the grid below. Be sure to label which perspective you are drawing.

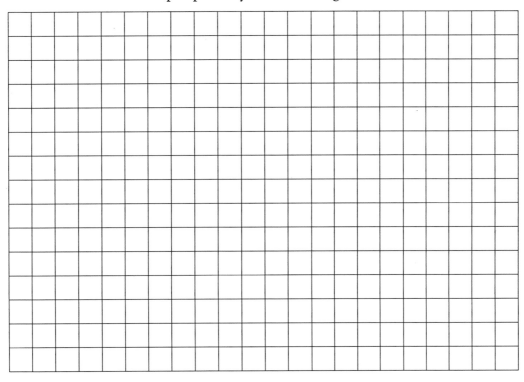

What were your strategies for this activity? _____

_____

_____

Can you see all the blocks from all perspectives? Explain. _____

_____

_____

Geometry
Set 2: Visualizing Solid Figures

Station 2

At this station you will be working with nets.

Below are possible nets for cubes.

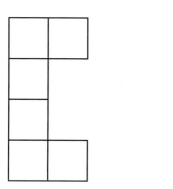

Could you fold this along the lines to make a cube? _____

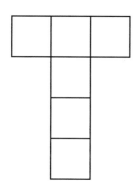

Could you fold this along the lines to make a cube? _____

Could you fold this along the lines to make a cube? _____

continued

Geometry
Set 2: Visualizing Solid Figures

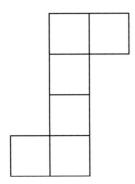

Could you fold this along the lines to make a cube? _____

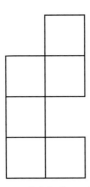

Could you fold this along the lines to make a cube? _____

Draw a net of your own that can be folded along the lines to make a cube.

Geometry
Set 2: Visualizing Solid Figures

Station 3

At this station, you will find 25 connecting cubes. You will use these to build a figure.

Below is a figure drawn on isometric dot paper. Make this figure using connecting cubes.

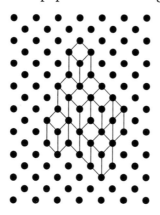

How many cubes did you use? _____

Are there any cubes that you cannot see in the drawing? Explain. _____

_____

_____

Now make a figure of your own out of the connecting cubes. Draw your figure.

What were your strategies for completing this task? _____

_____

_____

_____

Geometry
Set 2: Visualizing Solid Figures

Station 4

At this station, you will find two sheets of paper and tape. You will use these to draw nets for two cylinders.

Tape one sheet of paper into a cylinder so that 11 inches is the circumference. Tape the other sheet of paper into a cylinder so that 11 inches is the height.

Draw the net for each cylinder on the grid below. You may draw the net to be $\frac{1}{4}$ of the actual size. Be sure to label which sheets of paper belong to which cylinder.

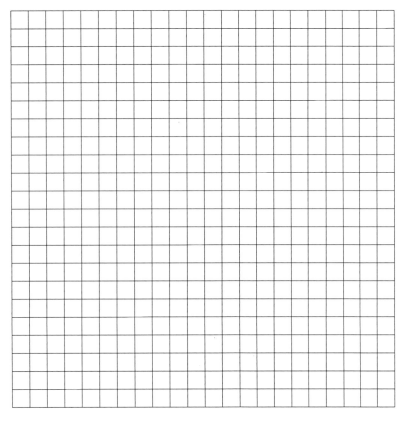

What similarities do you see between the two nets? _____

_____

_____

What differences do you see between the two nets? _____

_____

_____

Which cylinder has a greater surface area? _____

Geometry

Set 3: Problem Solving with Volume, Surface Area, and Scale

Goal: To provide opportunities for students to develop concepts and skills related to problem solving involving volume, surface area, and scaling

Common Core Standards

Geometry

Solve real-world and mathematical problems involving area, surface area, and volume.

6.G.1. Find the area of right triangles, other triangles, special quadrilaterals, and polygons by composing into rectangles or decomposing into triangles and other shapes; apply these techniques in the context of solving real-world and mathematical problems.

6.G.2. ... show that the volume is the same as would be found by multiplying the edge lengths of the prism. Apply the formulas $V = lwh$ and $V = bh$ to find volumes of right rectangular prisms with fractional edge lengths in the context of solving real-world and mathematical problems.

Student Activities Overview and Answer Key

Station 1

Students construct a cylinder out of a sheet of printer paper. They explore whether the way they choose to roll the paper affects the volume. They experiment and use calculations to answer this question.

Answers

Answers will vary; the cylinder with the 11-inch circumference; 257.1 cubic inches; 198.7 cubic inches; the cylinder with the 11-inch circumference

Station 2

At this station, students use the formula for the volume of a cone to determine how many ice cream cones could fit into a cooler. They explain how they do this, and are instructed to be careful with their units.

Answers

πr^2; 12.57 cubic inches; 12 cubic feet; 137 cones; you need to change feet to inches or inches to feet so you are working in the same unit, then you can divide the volume of a cone into the volume of the cooler

Station 3

Students use their knowledge of surface area to determine the surface area of a bird house. They need to be able to find the surface area of a rectangular prism and of a cylinder. Finally, they explain how they arrived at their solution.

Answers

128 square inches; 6.28 square inches—the area of the 2 circles; 6.28 square inches—the distance around the inside of the circle; 128 square inches; answers will vary

Station 4

Students use a photo and a picture frame to explore scale factor. They need to determine the scale factor necessary to enlarge the photo so that it fits in the frame.

Answers

Finding the dimensions to the photo and frame; answers will vary; answers will vary; answers will vary; find the ratio between the photo and the frame; ask them to enlarge the photo by the scale factor

Materials List/Setup

Station 1	two sheets of 8 ½ " × 11 " paper, tape, and mini marshmallows
Station 2	calculator
Station 3	calculators for all group members
Station 4	calculators and rulers for all group members a small picture and a larger picture frame (they must be similar)

Discussion Guide

To support students in reflecting on the activities, and to gather formative information about student learning, use the following prompts to facilitate a class discussion to "debrief" the station activities.

Prompts/Questions

1. Why does the way you roll a rectangle affect the volume of the cylinder you create?

2. What is a real-life example of when knowing the volume of an object is important?

3. What is a real-life example of when knowing the surface area of an object is important?

4. What is a real-life example of when scale factor is important?

Think, Pair, Share

Have students jot down their own responses to questions, discuss their responses with a partner (who was not in their station group), and then discuss as a whole class.

Suggested Appropriate Responses

1. One way, the length of the rectangle is the height of the cylinder, and the other way, it is the circumference of the cylinder.

2. Many examples—a can of soup, storage container, etc.

3. Many examples—when painting, wrapping, covering, etc.

4. Many examples—blueprints, maps, plans, etc.

Possible Misunderstandings/Mistakes

- Not being able to find radius when given the circumference

- Having trouble figuring out the surface area of the hole in the bird house

- Measuring the outside of the picture frame instead of the area that the picture will go in

Geometry
Set 3: Problem Solving with Volume, Surface Area, and Scale

Station 1

At this station, you will find two sheets of paper, tape, and mini marshmallows. Each sheet of paper is $8 \frac{1}{2}$" $\times$ 11". You will be using these to investigate volume.

Your task is to answer this question: Which way should you roll the paper into a cylinder to get the greatest volume? with 11 inches as the circumference or the height? Or, does it matter?

Tape one sheet of paper into a cylinder so that 11 inches is the circumference. Tape one sheet of paper into a cylinder so that 11 inches is the height.

Which do you think holds more? _____

Fill both cylinders with mini marshmallows.

Which cylinder holds more? _____

Now calculate the volume of the two cylinders. Write your solution below.

Volume with an 11" circumference: _____
Volume with an 11" height: _____

Which cylinder has a greater volume? _____

© 2011 Walch Education

Geometry
Set 3: Problem Solving with Volume, Surface Area, and Scale

Station 2

At this station, you will pretend you are an ice cream vendor. You have a cooler that is 3 feet × 2 feet × 2 feet. You want to fit in as many ice cream cones as possible. Each ice cream cone is pre-wrapped and has ice cream in it, so you cannot fit them inside one another.

The way to find the volume of a cone is the same as finding the volume of a pyramid, $\left(\dfrac{1}{3}\right)bh$ In the case of the cone, the base is not found by multiplying the length times the width.

How do we find the area of the base of a cone? _____

_____

_____

Each ice cream cone is 4 inches tall, and the radius is 1 inch.

What is the volume of one ice cream cone? _____

What is the volume of the cooler? _____

How many ice cream cones can you fit in the cooler? (Be careful of your units!)

_____

Explain how you arrived at your solution. _____

_____

_____

_____

Geometry

Set 3: Problem Solving with Volume, Surface Area, and Scale

Station 3

Imagine you wanted to build and paint a bird house. You have already completed most of the bird house, but you still have the most challenging piece to worry about: the front piece.

The front of the bird house looks like the drawing below. The radius of the circle is 1 inch. The block of wood is 1 inch by 4 inches by 12 inches.

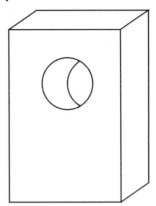

What is the surface area of the block of wood if it did not have the circle in it?

_____

What surface area are you losing because of the missing circle? Explain. _____

_____

_____

What surface area are you gaining because the circle is missing? Explain. _____

_____

_____

What is the overall surface area with the circle cut out? _____

Explain your strategy for figuring this out. _____

_____

_____

_____

Geometry
Set 3: Problem Solving with Volume, Surface Area, and Scale

Station 4

At this station, you will find enough calculators and rulers for all group members. You will also find a small picture and a larger picture frame.

You are asked to enlarge the small picture so it will fit into the picture frame.

What is the first step to solving this problem? _____

_____

_____

What are the dimensions of the photo?

        Length = _____

        Width = _____

        Height = _____

What are the dimensions of the frame?

        Length = _____

        Width = _____

        Height = _____

What is the scale factor? _____

How did you arrive at that answer? _____

_____

_____

Now that you have this information, what would you ask for if you went into a photo shop?

_____

_____

_____

Statistics and Probability

Set 1: Collecting, Organizing, and Analyzing Data

Goal: To provide opportunities for students to develop concepts and skills related to data collection and analysis

Common Core Standards

Statistics and Probability

Develop understanding of statistical variability.

6.SP.1. Recognize a statistical question as one that anticipates variability in the data related to the question and accounts for it in the answers.

6.SP.2. Understand that a set of data collected to answer a statistical question has a distribution, which can be described by its center, spread, and overall shape.

6.SP.3. Recognize that a measure of center for a numerical data set summarizes all of its values with a single number, while a measure of variation describes how its values vary with a single number.

Summarize and describe distributions.

Student Activities Overview and Answer Key

Station 1

Students will survey their classmates and turn that data into a stem-and-leaf plot. They will use the plot to analyze the data, and then reflect on how a stem-and-leaf plot is helpful.

Answers

Answers will vary; answers will vary; a stem-and-leaf plot helps us organize our data so it is easier to read.

Station 2

Students will be generating their own data and turning it into a scatter plot. They will examine and answer questions pertaining to this scatter plot.

Answers

Answers will vary.

Station 3

Students will formulate different questions for which they would be able to go out and collect data. They then choose one question to look at more closely. Students reflect on how they would work toward answering that question.

Answers

Answers will vary.

Station 4

Students try to answer the question of which comes up more, heads or tails, when flipping a coin. They conduct an experiment to help answer this question, then extend that to 10,000 flips.

Answers

Answers will vary; heads and tails should come up about the same number of times.

Materials List/Setup

Station 1   none

Station 2   ruler; two number cubes (1–6)

Station 3   none

Station 4   pennies for all group members

Discussion Guide

To support students in reflecting on the activities, and to gather formative information about student learning, use the following prompts to facilitate a class discussion to "debrief" the station activities.

Prompts/Questions

1. How does conducting an experiment help us answer questions?

2. What types of questions are best answered by numerical data?

3. What is an example of a real-life situation where we might want to conduct a survey to answer a question?

4. When would it be a good idea to use a stem-and-leaf plot in a real-world situation?

Think, Pair, Share

Have students jot down their own responses to questions, discuss their responses with a partner (who was not in their station group), and then discuss as a whole class.

Suggested Appropriate Responses

1. It lets us see how frequently we get different outcomes and what those outcomes are.

2. questions that ask how many

3. Anything where we care what a specific population does, wants, etc. For example, how much money do most people spend at lunch?

4. when you want to organize data numerically

Possible Misunderstandings/Mistakes

- Not putting data in order for the stem-and-leaf plot
- Not appropriately labeling the bar graph
- Asking questions that are too broad

Statistics and Probability
Set 1: Collecting, Organizing, and Analyzing Data

Station 1

You will be surveying the students in your class. Choose one of the following questions to use for your research.

- How many hours per week do you spend on homework?
- How many hours per week do you spend watching TV?
- How many hours per week do you spend listening to music?

All class members should answer the question that your group chose. Keep track of the answers in a list on the lines below.

_____

_____

_____

Use this data to construct a stem-and-leaf plot.

Which piece of data is the median number of the set? _____

Which number of hours is the mode? _____

How does a stem-and-leaf plot help you analyze data? _____

_____

_____

Statistics and Probability
Set 1: Collecting, Organizing, and Analyzing Data

Station 2

At this station, you will find two number cubes. You will be using these number cubes to generate data which you will then graph.

Roll the number cubes one at a time. Record the number that comes up on the first number cube. Roll the second number cube and record the number. Then record the sum of the two number cubes. Do this 10 times and record the data in the table below.

First number cube	Second number cube	Sum of number cubes

Using this data, create a scatter plot in the space provided. Be sure to title the scatter plot and use appropriate labels.

What is the range of the sums? _____

What do you notice about your scatter plot? Is there a general trend? _____

_____

_____

Statistics and Probability
Set 1: Collecting, Organizing, and Analyzing Data

Station 3

At this station, you will come up with different questions you could answer by collecting data.

Work with your group to come up with five questions you could answer by going out and collecting data (e.g., What is the average number of siblings students in our class have?).

_____

_____

_____

_____

_____

Choose one of those questions to focus on.

How would you gather data to answer your question? _____

_____

_____

What materials would you need to gather the data (e.g., number cubes, etc.)? _____

_____

_____

How would you present your data (e.g., bar graph, table, etc.)?

_____

_____

Statistics and Probability
Set 1: Collecting, Organizing, and Analyzing Data

Station 4

At this station, you will try to answer the question of which side of a penny comes up more, heads or tails.

Each member of your group should flip a penny 25 times, keeping track of how many times heads and tails come up.

	Tally	Final number
Heads		
Tails		

Combine your group's data in the table below.

	Total
Heads	
Tails	

Based on this data, which do you think has a better chance of coming up, heads or tails? Why?

_____

_____

_____

What do you think would happen if you flipped a penny 10,000 times? Do you think either heads or tails would come up more than the other? Why?

_____

_____

_____

Statistics and Probability

Set 2: Constructing Frequency Distributions

Goal: To provide opportunities for students to develop concepts and skills related to frequency distributions, frequency tables, and graphs

Common Core Standards

Statistics and Probability

Summarize and describe distributions.

6.SP.4. Display numerical data in plots on a number line, including dot plots, histograms, and box plots.

6.SP.5. Summarize numerical data sets in relation to their context, such as by:

    a. Reporting the number of observations.

    b. Describing the nature of the attribute under investigation, including how it was measured and its units of measurement.

Student Activities Overview and Answer Key

Station 1

Students will look at a newspaper article to determine letter frequency. They will count the number of each letter in an article. They will then use this information to construct a frequency table.

Answers

Table: answers will vary; answers will vary: yes or no; answers will vary: probably *e* or *t*; answers will vary: probably *q*, *x*, or *z*; The tally column makes it easier to add up all the times each letter appears in the article.

Station 2

Students will look at the average amount of monthly rainfall for New York City. They construct a line graph using this information, and then answer questions based on their line graph.

Answers

May and November; 1.1 inches; You can easily see the general change in average rainfall amounts by looking at the graph.

Station 3

Students roll a number cube and put their data into a frequency table. They use this data to answer appropriate questions and reflect on how the distribution table helped them.

Answers

Answers will vary; The Tally column means that you don't have to write down all the numbers you roll. Instead, you just put a mark, which means you rolled a given number.

Station 4

Students will work as a group to analyze a scatter plot. They will draw conclusions based on their observations, and use those conclusions to make general statements and predictions.

Answers

Number of hours spent studying and test scores; The more a student studies for the test, the higher that student's score; between a 65 and 70; We looked at where one half is on the scatter plot, and looked at the other points around that area.

Materials List/Setup

Station 1   newspaper article—150 words or less in length

Station 2   ruler

Station 3   number cube (1–6)

Station 4   none

Discussion Guide

To support students in reflecting on the activities, and to gather formative information about student learning, use the following prompts to facilitate a class discussion to "debrief" the station activities.

Prompts/Questions

1. How do frequency tables help us sort data?

2. What conclusions can you draw from a scatter plot if all the points are in a line? if the points are randomly distributed?

3. What is a good, real-life situation to model with a scatter plot?

4. When should a line graph be used?

Think, Pair, Share

Have students jot down their own responses to questions, discuss their responses with a partner (who was not in their station group), and then discuss as a whole class.

Suggested Appropriate Responses

1. We use the tally column to keep track of our data as we collect it, which is an organized strategy. They are also in numerical order, which makes it easier to read.

2. If the points are in a line, there is a correlation between the two variables. If the points are randomly distributed, one variable does not affect the other.

3. anything that involves two variables that have a relationship—one affects the other

4. when we are looking at a variable over time

Possible Misunderstandings/Mistakes

*   Losing track of which letters have been counted
*   Mislabeling the graph/axes
*   Having trouble creating a scale for numbers with decimal endings

Statistics and Probability
Set 2: Constructing Frequency Distributions

Station 1

At this station, you will be learning about the frequency that letters appear in a newspaper article. With your group, count the number of each letter in the article. Record your data in the table below. Place a mark for each letter in the column titled "Tally." Then write the total number of marks in the "Total number" column after you have finished counting all the letters.

Letter	Tally	Total number	Letter	Tally	Total number
A			N		
B			O		
C			P		
D			Q		
E			R		
F			S		
G			T		
H			U		
I			V		
J			W		
K			X		
L			Y		
M			Z		

E is the most frequently used letter in the English language. Do your results support this statement?

_____

What letter(s) appears the most frequently? _____

What letter(s) appears the least frequently? _____

How does the "Tally" column help you organize your data? _____

_____

© 2011 Walch Education

Statistics and Probability
Set 2: Constructing Frequency Distributions

Station 2

At this station, you will use a table to answer questions.

The average rainfall in New York City for each month is shown in the table below.

Month	Jan	Feb	Mar	Apr	May	June	July	Aug	Sept	Oct	Nov	Dec
Amount of rain (in.)	3.3	3.1	3.9	3.7	4.2	3.3	4.1	4.1	3.6	3.3	4.2	3.6

Source: http://www.worldclimate.com/cgi-bin/data.pl?ref=N40W073+2200+305801C

Construct a line graph using this information in the space below.

What month has the most rainfall on average? _____

What is the range of rainfall? _____

Why is a line graph an appropriate way to display this data? _____

_____

_____

Statistics and Probability
Set 2: Constructing Frequency Distributions

Station 3

At this station, you will find a number cube.

Roll the number cube 30 times. Record your data in the distribution table below.

Number	Tally	Total occurrences
1		
2		
3		
4		
5		
6		

What do you notice about the overall trend of the data? Did you roll about the same amount of each number?

_____

_____

_____

How does the "Tally" column help you organize your data?

_____

_____

_____

Statistics and Probability
Set 2: Constructing Frequency Distributions

Station 4

Discuss the following scatter plot with your group members. Then work together to answer the questions that follow.

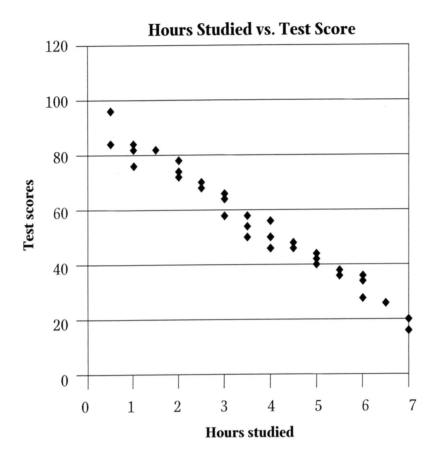

What two pieces of data are being compared in this scatter plot? _____

_____

What do you notice about the general trend of the data in this scatter plot? _____

_____

If you were told that a student spent half an hour studying, approximately what grade would you expect that student to earn on the test?

_____

What information did you use to make your last prediction? _____

_____

Statistics and Probability

Goal: To provide opportunities for students to develop concepts and skills related to using tables and graphs to examine variation in data

Common Core Standards

Statistics and Probability

Summarize and describe distributions.

6.SP.4.   Display numerical data in plots on a number line, including dot plots, histograms, and box plots.

6.SP.5.   Summarize numerical data sets in relation to their context, such as by:

    a.   Reporting the number of observations.

    b.   Describing the nature of the attribute under investigation, including how it was measured and its units of measurement.

    c.   Giving quantitative measures of center (median and/or mean) and variability (interquartile range and/or mean absolute deviation), as well as describing any overall pattern and any striking deviations from the overall pattern with reference to the context in which the data were gathered.

Student Activities Overview and Answer Key

Station 1

Students review the average monthly highs and lows for Monroe, Louisiana. They use this information to construct a double line graph. They then answer questions pertaining to their graph and reflect on why it is useful.

Answers

No, the lows will never be higher than the highs; the range; it allows us to see the differences between the highs and the lows and also see how the average temperature varies from month to month.

Station 2

Students will be generating their own data and turning it into a scatter plot. They will examine and answer questions pertaining to this scatter plot.

Answers

Answers will vary.

Station 3

Students will be looking at the make-up of packs of fun-size Skittles®. They will create a table using the Skittles®, and determine the variation in their data. Ultimately, they will draw conclusions about the make-up of a random pack of Skittles®.

Answers

Answers will vary; possibly that approximately the same number of each color is in each bag

Station 4

Students review the scores from Akron Acorns' games in February 2011. They then calculate the mean and range for the data, and answer questions based on these numbers. They look at the variation between the two teams.

Answers

2.6; 3.8; 5 goals; 4 goals; It looks like the Acorns lost more because their mean score is lower and they had more variation in their data. The opponents were consistently higher.

Materials List/Setup

Station 1 ruler

Station 2 ruler, two number cubes (1–6)

Station 3 fun-size packs of Skittles® (original fruit flavor) for every group member

Station 4 none

Discussion Guide

To support students in reflecting on the activities, and to gather formative information about student learning, use the following prompts to facilitate a class discussion to "debrief" the station activities.

Prompts/Questions

1. What is a real-life situation when you might want to know the range of data?

2. When is a double line graph a better choice than a scatter plot?

3. If the range in a set of data is very large, what does that say about the data?

4. What are some other ways to compare variation between two sets of data, besides looking at their means and range?

Think, Pair, Share

Have students jot down their own responses to questions, discuss their responses with a partner (who was not in their station group), and then discuss as a whole class.

Suggested Appropriate Responses

1. Possible answer: the cost of CDs at a store

2. when comparing two different values over time

3. There is a lot of variation; not consistent

4. looking at the medians

Possible Misunderstandings/Mistakes

- Not appropriately labeling the axes on graphs

- Having difficulty plotting two lines on the same graph

- Counting both end numbers in the range (e.g., smallest number 2, greatest number 6, and believing that since 2, 3, 4, 5, 6 is 5 numbers, the range is 5)

Statistics and Probability
Set 3: Using Tables and Graphs

Station 1

At this station, you will be constructing a double line graph.

Below are the average monthly highs and lows for Monroe, Louisiana.

Month	High	Low
January	56°F	33°F
February	61°F	37°F
March	69°F	45°F
April	76°F	53°F
May	84°F	61°F
June	91°F	69°F
July	94°F	72°F
August	94°F	70°F
September	89°F	64°F
October	79°F	51°F
November	68°F	43°F
December	59°F	36°F

Source: http://www.weather.com/weather/wxclimatology/monthly/graph/USLA0319

Use this data to construct a double line graph in the space below. Be sure to title the graph and use appropriate labels. Then answer the questions on the next page.

continued

Statistics and Probability
Set 3: Using Tables and Graphs

Do the two lines ever cross? Why or why not? _____

_____

_____

What does the distance between the two lines represent? _____

_____

_____

Why is a double line graph an appropriate way to display this data? _____

_____

_____

Statistics and Probability
Set 3: Using Tables and Graphs

Station 2

At this station, you will find two number cubes. You will be using these number cubes to generate data which you will then graph.

Roll the number cubes one at a time. Record the number that comes up on the first number cube. Roll the second number cube, and record the number. Then record the sum of the two number cubes. Do this 10 times, and record the data in the table below.

First number cube	Second number cube	Sum of number cubes

Using this data, create a scatter plot in the space below. Be sure to title the scatter plot and use appropriate labels.

What is the range of the sums? _____

What do you notice about your scatter plot? Is there a general trend? _____

_____

Statistics and Probability
Set 3: Using Tables and Graphs

Station 3

At this station, each group member will receive a fun-size bag of Skittles®. Do not eat these because you need to use them to find how much the contents vary between bags.

Each group member should open his/her bag of Skittles®. Record the number of each color of Skittle® in the table below.

Group member	Purple	Red	Yellow	Orange	Green
Total:					

What is the mean number of red Skittles® in each bag? _____

What is the overall mean number of Skittles® in each bag? _____

What is the range number of green Skittles®? _____

What is the range number when looking at the total for each color? _____

Are the totals about the same for each color? _____

What conclusions can you draw about the make-up of the average bag of fun-size Skittles®?

_____

_____

Statistics and Probability

Set 3: Using Tables and Graphs

Station 4

At this station, you will be looking at the scores from the Akron Acorns' games in February 2011. Below is a table with the results.

Akron Acorns' score	Opponents' score
2	5
2	5
3	4
6	3
2	3
1	5
1	4
3	5
3	1
4	5
1	4
3	2

What is the mean score for the Acorns? _____

What is the mean score for the opponents? _____

What is the range of scores for the Acorns? _____

What is the range of scores for the opponents? _____

Based only on your work above, do you think the opponents won or lost more games in February 2011? Why?

_____

_____

_____

Statistics and Probability

Goal: To provide opportunities for students to develop concepts and skills related to analyzing data using mean, median, and mode

Common Core Standards

Statistics and Probability

Develop understanding of statistical variability.

6.SP.2. Understand that a set of data collected to answer a statistical question has a distribution, which can be described by its center, spread, and overall shape.

6.SP.3. Recognize that a measure of center for a numerical data set summarizes all of its values with a single number, while a measure of variation describes how its values vary with a single number.

Student Activities Overview and Answer Key

Station 1

Students review a given set of index cards with numbers on them. They work as a group to determine the mean, median, mode, and possible outliers. Students decide which best represents the data. They then explain their strategy for determining which measure of central tendency is most representative of the set of data.

Answers

7.2; 5.5; 3; Yes, 27; median; possible explanation: We decided the median is most representative of the data because the 27 makes the mean too high and the mode is 3, which is too low. The median is in the middle and the 27 does not affect it, so we believe that is the best answer.

Station 2

Students will work with group members to construct three sets of data. A different measure of central tendency should be most representative of each set of data. Students will then reflect on their strategies for creating each set of data.

Answers

Answers will vary depending on data choice.

Possible strategies: either including or not including an outlier, having many of the same number in the data set, etc.

Station 3

Students randomly choose numbered index cards to create sets of data. They analyze the data to determine if there are any outliers present. They then reflect on how an outlier would affect the mean of a set of data.

Answers

Answers will vary depending on the index cards chosen. Yes, the mean will be affected by the outliers. The outliers will raise the overall mean.

Station 4

Students model mean, median, mode, and range by using data that they generate from number cubes. They work as a group to ensure that the range and central tendencies are modeled correctly. Students then reflect on their work.

Answers

Answers will vary—possible answer 1, 3, 4, 5, 2, 6, 3, 2, 1, 5; answers will vary—possible answer 1, 1, 2, 2, 3, 3, 4, 5, 5, 6; answers will vary and may range from 1 to 6; answers will vary and may range from 1 to 6; add the fifth and sixth number together then divide by 2, answers will vary and may range from 1 to 6; yes, if there is more than one number that is present most frequently

Materials List/Setup

Station 1	10 index cards with the following numbers written on them:
	2, 3, 3, 3, 5, 6, 7, 8, 8, 27
Station 2	none
Station 3	20 index cards with the following numbers written on them:
	1, 1, 1, 2, 2, 3, 3, 3, 4, 4, 4, 5, 6, 6, 7, 7, 7, 32, 45, 72
Station 4	number cube (numbered 1–6)

Discussion Guide

To support students in reflecting on the activities and to gather some formative information about student learning, use the following prompts to facilitate a class discussion to "debrief" the station activities.

Prompts/Questions

1. What is a real-life example of a set of data that would have an outlier?

2. Which of the central tendencies—mean, median, or mode—is not affected by outliers?

3. Can a set of data have more than one mean, median, or mode? If so, which?

4. Can the mean, median, and mode ever be the same number? Explain.

Think, Pair, Share

Have students jot down their own responses to questions, then discuss with a partner (who was not in their station group), and then discuss as a whole class.

Suggested Appropriate Responses

1. possibly the salaries at a company—the boss would make a lot more than the workers

2. median and mode

3. Yes. A set of data can have more than one mode.

4. Yes—if all the data is the same number, a set of data such as 1, 2, 3, 4, 5, 5, 5, 6, 7, 8, 9, etc.

Possible Misunderstandings/Mistakes

- Believing that the mean is always the best way to represent data
- Not recognizing outliers
- Believing that the mean, median, and mode must all be different numbers

Statistics and Probability

Set 4: Measures of Central Tendency

Station 1

Place the index cards you receive in order from least to greatest.

Work with your group to determine the mean of this data. Write your answer on the line below.

_____

Work with your group to determine the median of this data. Write your answer on the line below.

_____

Work with your group to determine the mode of this data. Write your answer on the line below.

_____

Are there any outliers in this set of data? If so, what number?

_____

Which one of the measures of central tendency—mean, median, or mode—do you believe is the best representation of this set of data?

_____

Explain how you made your decision and why you believe it is the right decision.

_____

_____

_____

Statistics and Probability
Set 4: Measures of Central Tendency

Station 2

At this station, your group will be creating data of your own. Your goal is to have three sets of data—one where the mean best represents the data, one where median best represents the data, and one where the mode best represents the data. Each set of data should have five numbers. Discuss each measure and record your numbers after everyone agrees.

Write five numbers in which the mean is the best measure of the data.

☐ ☐ ☐ ☐ ☐

What is the mean of this set of data? _____

Write five numbers in which the median is the best measure of the data.

☐ ☐ ☐ ☐ ☐

What is the median of this set of data? _____

Write five numbers in which the mode is the best measure of the data.

☐ ☐ ☐ ☐ ☐

What is the mode of this set of data? _____

What strategies did you use to help come up with these sets of data? _____

_____

_____

_____

_____

Statistics and Probability
Set 4: Measures of Central Tendency

Station 3

At this station, your group will be given a set of index cards. Shuffle these cards and place them face down. Choose seven cards without looking and turn them over.

Record your cards below.

☐ ☐ ☐ ☐ ☐ ☐ ☐

Does your set of data have an outlier? If so, what was it? _____

Repeat this two more times.

☐ ☐ ☐ ☐ ☐ ☐ ☐

Does this set of data have an outlier? If so, what was it? _____

☐ ☐ ☐ ☐ ☐ ☐ ☐

Does this set of data have an outlier? If so, what was it? _____

Now look through all the index cards. Do you think the outliers would affect the mean of the data? Explain.

_____

_____

_____

Statistics and Probability
Set 4: Measures of Central Tendency

Station 4

You will find a number cube at this station. Use the number cube to create a set of data. Work as a group to determine the mean, median, and mode of data that you generated.

Roll the number cube 10 times. Record your results in the boxes below.

<table>
<tr><td> </td><td> </td><td> </td><td> </td><td> </td><td> </td><td> </td><td> </td><td> </td><td> </td></tr>
</table>

Rearrange the numbers in the boxes above so they are in order from least to greatest. Record these numbers in the boxes below.

<table>
<tr><td> </td><td> </td><td> </td><td> </td><td> </td><td> </td><td> </td><td> </td><td> </td><td> </td></tr>
</table>

Work with other students to calculate the mean of this data. Write your answer in the space below and explain how you got the mean.

_____

_____

_____

Work with other students to decide the median of this data. Write your answer in the space below and explain how you got the median.

_____

_____

_____

continued

Statistics and Probability
Set 4: Measures of Central Tendency

Explain how you would find the median if your fifth and sixth piece of data are not the same number.

_____

_____

_____

Work with other students to identify the mode of this data. Write your answer in the space below and explain how you got the mode.

_____

_____

_____

Is it possible to have more than one mode? If so, when does that happen? Explain your answer below.

_____

_____

_____

Statistics and Probability

Set 5: Measures of Variation

Goal: To provide opportunities for students to develop concepts and skills related to measures of variation in data

Common Core Standards

Statistics and Probability

Develop understanding of statistical variability.

6.SP.2. Understand that a set of data collected to answer a statistical question has a distribution, which can be described by its center, spread, and overall shape.

6.SP.3. Recognize that a measure of center for a numerical data set summarizes all of its values with a single number, while a measure of variation describes how its values vary with a single number.

Student Activities Overview and Answer Key

Station 1

Students review the average monthly precipitation for a town in the northeast. They then determine the values of the first and third quartile and the median. Students reflect on what these values tell them about the data.

Answers

4.3; 3.65; 4.8; They give you a general idea about the variation in the data.

Station 2

Students will explore the variation of colors in fun-size bags of M&Ms®. They will count the number of each color, and use these numbers to comment on the overall variation in color.

Answers

Answers will vary.

Station 3

Students will look at data from the Newtown Hawks in February 2011. They will find the interquartile range of this data, and draw conclusions about whether the Hawks had a winning month.

Answers

96; 103; 20; 8; answers may vary—the Hawks did not perform better than their opponents. Their median score is lower and they have a lot more variation in their score.

Station 4

Students will collect data from rolling number cubes. They will then construct two box-and-whisker plots. They will compare the variation between their two sets of data.

Answers

Answers will vary; answers will vary; answers will vary—There are more possible values for the sum so there should be more variation.

Materials List/Setup

Station 1	none
Station 2	a fun-size bag of M&Ms® for each student
Station 3	none
Station 4	two number cubes (1–6)

Discussion Guide

To support students in reflecting on the activities and to gather some formative information about student learning, use the following prompts to facilitate a class discussion to "debrief" the station activities.

Prompts/Questions

1. In what real-life situation might it be important to know the variation within a set of data?

2. In what real-life situation might it be important to know the variation between two sets of data?

3. Why is it important that we identify outliers in our data?

4. Why is interquartile range a useful number for use to know about a set of data?

Think, Pair, Share

Have students jot down their own responses to questions, then discuss with a partner (who was not in their station group), and then discuss as a whole class.

Suggested Appropriate Responses

1. Possible answer—the prices of DVDs at a store

2. Possible answer—when comparing the price of DVDs at two stores

3. They skew our results—make the whiskers longer than they should be

4. It tells us about the middle 50% of the data.

Possible Misunderstandings/Mistakes

- Not understanding how to find a quartile value if it is between two data points
- Not identifying outliers in the data
- Confusing the first quartile with the third quartile

Statistics and Probability
Set 5: Measures of Variation

Station 1

At this station, your group will be looking at the average monthly precipitation for a town in the northeast. Below is a table with this information.

Month	Jan	Feb	Mar	Apr	May	June	July	Aug	Sept	Oct	Nov	Dec
Precipitation (inches)	4.8	4.8	5.8	4.3	4.3	3.6	5.0	3.7	3.4	3.1	3.9	4.3

What is the median amount of precipitation? _____

What is the first quartile value? _____

What is the third quartile value? _____

What do these values tell you about the average monthly precipitation in this town?

_____

_____

_____

Statistics and Probability
Set 5: Measures of Variation

Station 2

At this station, each group member will receive a fun-size bag of M&Ms®. You may not eat these because you will use them to help find how much the contents vary between bags.

Each group member should open his/her bag of M&Ms. Record the number of each color M&M in the table below.

Group member	Brown	Red	Yellow	Orange	Green	Blue
Total						

What is the range of green M&Ms? _____

What is the range when looking at the totals for each color? _____

What is the median when looking at the totals for each color? _____

Are there any outliers when looking at the totals for each color? _____

What conclusions can you draw about the variation of the average bag of fun-size M&Ms?

_____

_____

_____

Statistics and Probability
Set 5: Measures of Variation

Station 3

At this station, your group will be looking at the scores of the Newtown Hawks for February 2011. Below is a table with the scores from each game.

Newtown Hawks	Opponent	Newtown Hawks	Opponent
91	103	96	93
101	99	81	106
83	90	96	103
5	87	105	99
120	111	106	115
106	105	87	110
76	102	100	107

What is the median score for the Newtown Hawks? _____

What is the median score for the opponents? _____

What is the interquartile range for the Newtown Hawks? _____

What is the interquartile range for the opponents? _____

What does this information tell you about the Newtown Hawks? Did they perform better than their opponents in general? Explain.

_____

_____

_____

Statistics and Probability
Set 5: Measures of Variation

Station 4

At this station, you will find two number cubes. Your group will be using these number cubes to generate data which you will then graph.

Roll the number cubes one at a time. Record the number that comes up on the first number cube. Roll the second number cube and record. Then record the sum of the two number cubes. Do this ten times and record the data in the table below.

First number cube	Second number cube	Sum of number cubes

Using this data, create two box-and-whisker plots, one for the first column of data and one for the third column of data. It may be helpful to split the group in half, each half working on one box-and-whisker plot.

What is the interquartile range for one number cube? _____

What is the interquartile range for the sum of the number cubes? _____

Explain why there might be a difference in the variation. _____

_____

_____

_____

Statistics and Probability

Goal: To provide opportunities for students to develop concepts and skills related to analyzing data using appropriate graphs

Common Core Standards

Statistics and Probability

Summarize and describe distributions.

6.SP.4. Display numerical data in plots on a number line, including dot plots, histograms, and box plots.

6.SP.5. Summarize numerical data sets in relation to their context, such as by:

    a. Reporting the number of observations.

    b. Describing the nature of the attribute under investigation, including how it was measured and its units of measurement.

    c. Giving quantitative measures of center (median and/or mean) and variability (interquartile range and/or mean absolute deviation), as well as describing any overall pattern and any striking deviations from the overall pattern with reference to the context in which the data were gathered.

Student Activities Overview and Answer Key

Station 1

Students will work as a group to analyze a scatter plot. They will draw conclusions based on their observations and use those conclusions to make general statements and predictions.

Answers

Number of hours spent exercising and percentage of body fat; The more a person exercises, the lower that person's percentage of body fat; between 40% and 50%; We looked at where one half is on the scatter plot and looked at the other points around that area.

Station 2

Students will construct a box-and-whisker plot. They will use this to answer questions about the data.

Answers

87.5; 93; yes, 60; A box-and-whisker plot helps us see the how spread out our data is.

Station 3

Students will survey their group members to obtain data. They will construct a circle graph using this data. They then reflect on the experience and evaluate the choice to use a circle graph for this exercise.

Answers

Table 1—answers will vary; Table 2—answers will vary based on the results in Table 1; circle graph—answers will vary based on the results in Table 2; Seeing our data in a circle graph helps us to see how long most people spend doing their activity and helps us see which range of time is most popular.

Station 4

Students will work as a group to make two different histograms representing the same data. They explore how the interval of a graph affects the appearance of the graph and answer questions associated with that concept.

Answers

The two graphs have the same information but the smaller the interval, the more informative the graph is; Yes. You can see the breakdown of grades better with the 5-point interval.

Materials List/Setup

Station 1   none

Station 2   ruler; calculator

Station 3   ruler; calculator; protractor

Station 4   ruler

Discussion Guide

To support students in reflecting on the activities and to gather some formative information about student learning, use the following prompts to facilitate a class discussion to "debrief" the station activities.

Prompts/Questions

1. When would a circle graph be the best choice to display information?

2. How can the interval we choose for a graph change the way we perceive the data being displayed by the graph?

3. What conclusions can you draw from a scatter plot if all the points are in a line? What if the points are randomly distributed?

4. What is a good real-life situation to model with a scatter plot?

Think, Pair, Share

Have students jot down their own responses to questions, then discuss with a partner (who was not in their station group), and then discuss as a whole class.

Suggested Appropriate Responses

1. when you are interested in percents or making a visual comparison of data

2. If an interval is too large, we do not see the details of the data.

3. If the points are in a line, there is a correlation between the two variables. If the points are randomly distributed, one variable does not affect the other.

4. Anything that involves two variables that have a relationship—one affects the other.

Possible Misunderstandings/Mistakes

- Not understanding how to properly label graphs
- Confusing axes
- Having trouble converting percents to degrees

NAME: _____

Statistics and Probability
Set 6: Analyzing Data Using Graphs

Station 1

Discuss the following scatter plot with your group members, and then work together to answer the questions that follow.

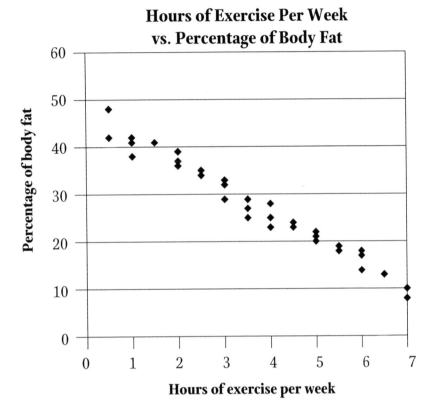

What two pieces of data are being compared in this scatter plot? _____

_____

What do you notice about the general trend of the data in this scatter plot? _____

_____

If you were told that a person spent less than half an hour exercising per week, about what percentage of body fat would you expect that person to have? _____

What information did you use to make that prediction? _____

_____

    Station Activities for Common Core Mathematics, Grade 6

Statistics and Probability
Set 6: Analyzing Data Using Graphs

Station 2

At this station, your group will construct a box-and-whisker plot to analyze the scores students earned on a test.

On a recent math test, 10 students earned the following scores:

Scores
60
84
100
78
86
89
92
75
95
93

Use this information to construct a box-and-whisker plot in the space below.

What is the median of this set of data? _____

What is the third quartile? _____

Are there any outliers in this set of data? If so, what are they? _____

How does a box-and-whisker plot help you understand data? _____

_____

_____

Statistics and Probability
Set 6: Analyzing Data Using Graphs

Station 3

You will be surveying your group members. Choose one of the following questions that you will use for your research.

1. How many hours per week do you spend on homework?

2. How many hours per week do you spend watching television?

3. How many hours per week do you spend listening to music?

All group members should answer the question that your group chose. Keep track of the answers in the table below. If a group member's answer is on the border (e.g., 6 hours), use the higher range (e.g., 6–9 hours).

Answer	Tally	Total number
0–3 hours		
3–6 hours		
6–9 hours		
9–12 hours		
12–15 hours		
15 or more hours		

As a group, use scrap paper and a calculator to determine the percent of the total number of people that each category represents. Fill in this information in the table below. Use this information to help you determine how many degrees out of 360 each category would fill in a circle graph.

Answer	Percent	Number of degrees out of 360
0–3 hours		
3–6 hours		
6–9 hours		
9–12 hours		
12–15 hours		
15 or more hours		

continued

Statistics and Probability

Set 6: Analyzing Data Using Graphs

Use this information, a protractor, and a ruler to construct a circle graph in the space below.

How is viewing this information in a circle graph useful? _____

_____

_____

Statistics and Probability
Set 6: Analyzing Data Using Graphs

Station 4

At this station, your group will be constructing two different histograms. Use the following set of data for both histograms.

A teacher gives a test to her twenty students. They received the following grades:

87, 80, 90, 60, 78, 79, 80, 95, 97, 98, 99, 95, 75, 67, 79, 83, 84, 100, 97, 83

For the first histogram, the intervals for grades should be ten points. Construct this histogram in the space below.

For the second histogram, the intervals for grades should be five points. Construct this histogram in the space below.

Discuss the histograms with your group. What do you notice about the two graphs? Is one more informative than the other?

_____

_____

How does having different intervals affect how the graph looks? Explain. _____

_____

_____